Turbulence Nature and the Inverse Problem

# FLUID MECHANICS AND ITS APPLICATIONS
Volume 89

*Series Editor:* R. MOREAU
MADYLAM
*Ecole Nationale Supérieure d'Hydraulique de Grenoble*
*Boîte Postale 95*
*38402 Saint Martin d'Hères Cedex, France*

*Aims and Scope of the Series*

The purpose of this series is to focus on subjects in which fluid mechanics plays a fundamental role.

As well as the more traditional applications of aeronautics, hydraulics, heat and mass transfer etc., books will be published dealing with topics which are currently in a state of rapid development, such as turbulence, suspensions and multiphase fluids, super and hypersonic flows and numerical modeling techniques.

It is a widely held view that it is the interdisciplinary subjects that will receive intense scientific attention, bringing them to the forefront of technological advancement. Fluids have the ability to transport matter and its properties as well as to transmit force, therefore fluid mechanics is a subject that is particularly open to cross fertilization with other sciences and disciplines of engineering. The subject of fluid mechanics will be highly relevant in domains such as chemical, metallurgical, biological and ecological engineering. This series is particularly open to such new multidisciplinary domains.

The median level of presentation is the first year graduate student. Some texts are monographs defining the current state of a field; others are accessible to final year undergraduates; but essentially the emphasis is on readability and clarity.

For other titles published in this series, go to
www.springer.com/series/5980

L.N. Pyatnitsky

# Turbulence Nature and the Inverse Problem

With 75 Figures

Springer

L.N. Pyatnitsky
Russian Academy of Sciences (RAS)
Inst. Physicotechnical Problems in
Power Engineering
Panferov Street 8–139
Moskva, Moscow, Russian Federation
piat@sci.lebedev.ru

This is a completely revised and updated translation of the original Russian work "Navier-Stokes Equation and Turbulent Pulsations" (in Russian); Moscow, 2006, Russian Foundation for Basic Research.

ISBN: 978-90-481-8480-4 e-ISBN: 978-90-481-2251-6



Printed on acid-free paper

9 8 7 6 5 4 3 2 1

springer.com

# Preface

Turbulent dynamics of fluids and gases occurs in a variety of natural and technical systems. Turbulence plays a fundamental role in the atmospheres of the Earth and the Sun, the intermixing of fluids in oceans, and river flows. Understanding turbulence is necessary to calculate the parameters of media flows in pipes, flows of air around cars, aircrafts, and space vehicles in the atmosphere.

Due to the widespread occurrence of turbulence and the profound analogy between it and a range of other physical phenomena of statistical character turbulence studies are important not only from the standpoint of applications but also from that of basic science. Though turbulence has been investigated for over a century, the phenomenon is extremely complex and there still exists no general theory to describe it.

As various applications of turbulence necessitated developing methods of its simulation, experimental studies were performed and semi-empirical techniques of calculating the parameters of turbulent flows were developed on their basis. In the more recent past, the techniques materialized in the form of computer codes which made it much easier to solve a number of practical problems. However, the scope of the empirical methods is limited, and the task of creating a theory of turbulence remains high on the scientific agenda.

The nature of the oscillations of flow parameters and the mechanism of their emergence must be understood precisely to develop the fundamentals of turbulence theory. It should be noted that the issues have been considered already by Lord Rayleigh, Osborne Reynolds and also by Hendrick Lorentz whose concepts of the origin of turbulence diverged. An attempt to find a solution to the two basic problems – the nature and the mechanism of the emergence of turbulent oscillations – using the wave approach is made in the present book. The solution was found largely by chance during an investigation into the transition from normal combustion to detonation.

The evolution of the process in a pipe having a square cross-section was recorded by Schlieren technique. It transpired that the gas flow in front of the flame involved numerous perturbations looking like compression waves propagating in various directions with the speed of sound. The perturbations emerged at the pipe walls and represented spherical wave packets. In the course of time, the waves formed an increasingly dense three-dimensional system of perturbations, and the smooth flame turned turbulent.

Measurements showed that the pressure oscillations replicated those of the velocity. The examination of hundreds of snapshots and oscillograms led to the idea that the field of turbulent fluctuations was actually of acoustic nature and was formed by acoustic waves. The idea, however, collides with the broadly adopted notion that pressure in the transverse cross-section of the flow is constant. At that time the new concept raised serious doubts, especially since the emergence of perturbation waves at a pipe smooth wall also appeared strange. Therefore the theme required further investigation.

The process took decades. In the meantime I worked on the problem of neutralizing diffraction divergence to propagate optical beams over distances spanning many Rayleigh ranges. By the time the problem was solved I realized that the gas flow in a pipe should be regarded as a wave beam. According to Rayleigh's law elementary plane waves in a beam rapidly grow spherical. In the gas flow case the spherical waves are reflected by pipe walls and thus the conditions get formed for the generation of perturbations where the reflection occurs.

The next problem was to model specific phenomena numerically and to formulate a unified approach. *Mathematica* interactive computation system (Wolfram Research Inc.) was used for the purpose. *Mathematica* codes for the simulation of certain phases of the above process are presented below.

The topic of a large part of the book is the development of the model of turbulent oscillations. Calculation methods and relations describing the field of turbulent oscillations in space and time are presented. They are based entirely on the conservation laws. No new formulae are nearly needed to describe the process – all the expressions used below are taken from "Fluid Mechanics" and "The Classical Theory of Fields" by L.D. Landau and E.M. Lifshitz, the author's teachers. Only the interpretations and the applications of the corresponding formulae had to be adjusted for the purposes of the present study.

The wave model and the analytical representation of the spatio-temporal field of parameter oscillations open unique opportunities to explore theoretically various properties of the field and the way they are affected by the flow conditions. In particular temporal and spatial spectra of the oscillations are considered to that end. The chaotization of the oscillations of the flow parameters is a subject of a separate Chapter. In it the dependence of the extent of randomization on the character of the generation of perturbations and the conditions under which the oscillations can be treated as a purely random process are examined.

The experimental validation of the foundations of the wave model is also discussed in detail. Experimental evidence is provided to support the concepts of the flow structure in a simple wave and of the mechanism of the emergence of perturbations at the pipe walls. Calculated distributions of the longitudinal and transverse oscillation components agree well with those observed experimentally.

The transition from normal combustion to detonation provides an example of the application of the wave model of turbulent oscillations. Like turbulence, the phenomenon had been studied for a long time but its picture had not been finalized. Yet, recurrent blasts in coal mines make it necessary to solve the problem.

The analytical representation of the spatio-temporal field of turbulent oscillations makes it possible to move on to the specifics of the inverse problem of turbulence, by which I mean calculating the parameters of the original perturbations from the spatio-temporal oscillations data. The idea of posing and solving the inverse problem is applied below to calculate the parameters of an optical discharge caused by a laser beam with compensated diffraction divergence. The parameters of the divergence primary breakdowns that remained out of reach for experimental measurements have been calculated by solving the corresponding inverse problem.

The calculations of the spatio-temporal field of turbulent oscillations have been carried out for slow flows in channels so that the fluid compressibility can be ignored. It is planned to incorporate fluid compressibility and a number of other factors into the considerations in the future.

Moscow
December 2008

Lev N. Pyatnitsky

# Acknowledgments

I am indebted to late Professor K.I. Shchelkin and to late Professor M.D. Millionshchikov for encouraging me to pursue the turbulence model even at the time when the task seemed hopeless. I am delighted to acknowledge Professor V.M. Batenin's making it possible for me to carry out the work and Professor G.I. Barenblatt's getting into details of my views on the subject and pointing to the possibility of two parallel mechanisms responsible for the formation of turbulent oscillations. Thanks are also due to Professor H.M. Milchberg for providing an opportunity to carry out experiments with a short-pulse laser at his laboratory.

# Symbols

| | |
|---|---|
| $\boldsymbol{U}_f$, $P_f$, $\varrho_f$ | instantaneous values of fluid velocity, pressure and density |
| $\boldsymbol{U}$ | vector of mean fluid velocity |
| $U,V,W$ | components of velocity vector $\boldsymbol{U}$ |
| $M$ | Mach number |
| $\boldsymbol{u}$ | vector of velocity variations, velocity of oscillations |
| $u,v,w$ | components of velocity variations $\boldsymbol{u}$ and velocity oscillations |
| $\varphi$ | velocity potential |
| $P$ | fluid mean pressure |
| $P$ | pressure variations |
| $\varrho$ | fluid mean density |
| $\rho$ | density variations |
| $T$ | temperature, full duration of a process |
| $T_B$ | temperature of explosive mixture spontaneous ignition |
| Re | Reynolds number |
| $\mathrm{Re}_{cr}$ | Reynolds critical number |
| $\mathrm{Re}_x$ | local value of Reynolds number |
| $D$ | pipe diameter, velocity of shock and detonation waves |
| $d$ | characteristic dimension of channel, diameter of perturbation |
| $R$ | pipe radius, perturbation external radius |
| $r$ | instantaneous distance, radius |
| $\boldsymbol{r}(x,y,z)$ | instantaneous distance to a point of flow |
| $\vartheta$, $\varphi$ | angular coordinates |
| $\boldsymbol{\rho}$ | radius-vector of wave front |
| $\xi,\eta,\zeta$ | components of radius-vector, variable coordinates of integration |
| $\boldsymbol{r}_0$ | vector of observation point at coordinates $x_0,y_0,z_0$ |
| $\boldsymbol{rp}$ | vector of perturbation origin location at coordinates $xp,yp,zp$ (of one or sequence) |
| $tp$ | instant of perturbation origin (of one or sequence) |
| $n$ | number of wave reflections from channel walls |
| $L$ | length of simple wave, mixing path length |
| $L$ | length of acoustic radiation beam |
| $L$ | length of double saltus |

| | |
|---|---|
| $\ell$ | trajectory length of a probe point of acoustic wave |
| $\boldsymbol{s}$ | vector tangent to probe point trajectory |
| $t,\tau$ | time in measurement of various processes |
| $a$ | primary perturbation radial dimension, fluid thermal diffusivity factor |
| $f(R,c\tau)$ | primary distribution in perturbation |
| $A$ | oscillation amplitude |
| $\gamma$ | attenuation exponent of oscillation amplitude |
| $c$ | sound velocity, light velocity |
| $P$ | wave parameter in acoustics |
| $\boldsymbol{k}$ | wave vector |
| $k$ | wave number |
| $k_0$ | wave number in radiation beam axial direction |
| $k(\theta)$ | wave number in direction of angle θ |
| $\lambda$ | radiation wave-length, molecule free path |
| $\omega$ | oscillation of angular frequency |
| $\beta$ | arbitrary initial phase of a wave |
| $\alpha$ | Landau's constant |
| $\kappa$ | von Karman constant |
| $\psi_0$ | stream function |
| $\psi$ | disturbance stream function |
| $f(\zeta)$ | dimensionless stream function of dimensionless coordinate ζ |
| $\partial\omega/\partial\boldsymbol{k}$ | wave packet group velocity |
| $\Pi_{ik}$ | tensor of momentum flux density |
| $\sigma_{ik}$ | stress tensor |
| $i,j,k$ | component index |
| $\sigma$ | friction tangential stress |
| $\boldsymbol{J}$ | vector of momentum flux density of fluid |
| $\upsilon$ | molecule thermal velocity |
| $c_p/c_v$ | ratio of specific heats |
| $\eta$ | dynamic viscosity factor |
| $\nu$ | kinematic viscosity factor |
| $a$ | thermal diffusivity |
| $\delta$ | boundary layer thickness |
| $\delta_l$ | thickness of laminar boundary layer |
| $\delta_t$ | thickness of turbulent boundary layer |
| $\Delta$ | displacement of flame leading point |
| $A_t$ | factor of turbulent exchange |
| $\overline{u_i u_j}$ | pair correlation functions |
| $S$ | surface area |
| $F$ | pipe cross-section area, correlation function |
| $G(t,r)$ | function |
| G | region of Green function definition |

| | |
|---|---|
| $q$ | turbulent energy dissipation |
| $C$ | Kolmogorov's constant |
| $h$ | exponent of invariance scaling |
| $q_l$ | energy flux at scale $l$ |
| $q$ | specific energy flux from external source |
| $\Pi_{ik}$ | momentum flux density (1.12) |
| $\varepsilon$ | fluid internal energy per unit mass |
| $\varepsilon(k)$ | fluctuation kinetic energy |
| $H$ | fluid enthalpy |
| s | entropy per unit mass |
| $\Gamma(\boldsymbol{U})$ | velocity circulation |
| $E$ | kinetic energy of moving fluid |
| $\gamma$ | gas adiabatic exponent |
| $X$ | piston motion law |
| $X'$ | piston velocity |
| g | amplitude of piston velocity alteration |
| $\mu$ | exponent of piston motion law |
| $\theta$ | diffraction divergence angle |
| $\gamma$ | part of acoustic wave energy dissipated at reflection |
| $\gamma$ | intermittency factor |
| $\omega(\boldsymbol{k})$ | dispersion relation |
| $\delta(\omega)$ | Dirac delta function |
| $G_{\mathrm{n}}(r)$ | Gauss type function at exponent factor n |
| $J_{\mathrm{i}}(x)$ | Bessel function of the first order at index i |
| $x$ | Bessel function argument, $x = kr\mathrm{sun}\gamma$ |
| $E$ | electric field of optical radiation |
| $E_*$ | optical field critical value for self-focusing |

# Contents

1 **The turbulence problem** .......... **1**
1.1 The first interpretation .......... 1
1.2 The next approaches .......... 5
1.3 A new approach .......... 11

2 **Fluid motion** .......... **17**
2.1 Equations of fluid motion .......... 17
2.2 Vorticity .......... 20
2.3 Wave equation and incompressibility conditions .......... 23

3 **Distribution of parameters in viscous flow** .......... **29**
3.1 Velocity profiles in a flow cross-section .......... 29
3.2 Hypothesis on pressure profile in a flow cross-section .......... 33
3.3 Correction of the pressure profile .......... 38

4 **Perturbations in viscous flow** .......... **43**
4.1 Fluid motion from the start .......... 43
4.2 Simple wave and wave beam .......... 47
4.3 Origin of pressure perturbations .......... 51

5 **Perturbation in channels** .......... **55**
5.1 Perturbations in semi-infinite space .......... 55
5.2 Perturbation waves in flow .......... 58
5.3 Distortion of the wave packet in channels .......... 61
5.4 The wave packet in the boundary layer .......... 66

6 **Spatio-temporal field of perturbations in channels** .......... **73**
6.1 Computing technique of wave configuration in channels .......... 73
6.2 Wave front configuration appearance in channels .......... 76
6.3 Structure of flow perturbations in channels .......... 79

7 **Evolution of velocity oscillation field** .......... **85**
7.1 Oscillations of flow parameters produced by a wave .......... 85
7.2 Spatio-temporal field of oscillations in a wave sequence .......... 93
7.3 Chaotization of a spatio-temporal field .......... 97

**8 Experimental substantiation of turbulence wave model ........ 103**
8.1 Structure of a simple wave ........ 103
8.2 Boundary layer separation and flow perturbations ........ 114
8.3 Distribution of oscillations in flow cross-section ........ 119

**9 Transition from normal combustion to detonation ........ 125**
9.1 Short history of the problem ........ 125
9.2 Exposition of flame propagation in a pipe ........ 127
9.3 Initial stage of the flame propagation ........ 131
9.4 Uniform flame propagation and second acceleration ........ 135
9.5 Formation of detonation wave ........ 143

**10 An inverse problem of turbulence ........ 155**
10.1 Object of the inverse problem application ........ 155
10.2 Wave beam at Rayleigh divergence compensated ........ 158
10.3 Structures of plasma channels in lengthy wave beams ........ 160
10.4 Breakdown structures in the short heating impulse ........ 166
10.5 Formation of complex structures of the plasma channel ........ 171

**Conclusion ........ 181**
**References ........ 185**
**Index ........ 193**

# 1
# The turbulence problem

## 1.1 The first interpretation

The system of hydrodynamic equations including Navier–Stokes and continuity equations describes distributions of flow parameters, i.e., velocity $\boldsymbol{U}_f$, pressure $P_f$, density $\varrho_f$, and temperature $T_f$. However, O. Reynolds [1] observing liquid flows in a pipe at various velocities discovered a curious phenomenon. Some unknown forces appeared at certain liquid velocity to perturb the flow, and produce the velocity fluctuations. As a result, the liquid motion in the pipe became chaotic, or "turbulent". It was found later that all the other flow parameters undergo similar fluctuations.

The moving fluid state is descried by means of the fluid's velocity and any two thermodynamic quantities pertaining to the fluid, say, pressure and density. The fluctuations depend on flow conditions in more subtle line. However, numerous experiments testify that a parameter fluctuation amplitude makes up only a small part of the parameter mean quantity, which does not exceed some percents (see, e.g. [2]). Hence any of the turbulent flow parameters can be considered as a sum of a time-averaged basic value, and some small perturbation. If the time-averaged basic values are the velocity $\boldsymbol{U}(U, V, W)$, pressure $P$, density $\varrho$, and their small perturbations are accordingly $\boldsymbol{u}(u, v, w)$, $p$, $\rho$, then the relations can be written as

$$\boldsymbol{U}_f = \boldsymbol{U} + \boldsymbol{u},\ P_f = P + p,\ \varrho_f = \varrho + \rho. \tag{1.1}$$

Averaged components of the parameters can be computed by solving a system of hydrodynamic equations, at least in simple cases. However with regard to fluctuations and an impressive picture of a flow transition from the laminar state into the turbulent, the problem has not yet been solved. The state of the theory at this point has been emphasized by L.D. Landau and E.M. Lifshitz: "No complete quantitative theory of turbulence has yet been evolved" ([3], §33), or "a complete theory of turbulence (which does not yet exist) …" ([3], §35). The cause of such a strange situation lies in lack of understanding of the mechanism of turbulent fluctuations. This lack,



L.N. Pyatnitsky, *Turbulence Nature and the Inverse Problem*, Fluid Mechanics and Its Applications 89, 1–15

in its turn, means that the nature of primary exciting motion is still in abeyance.

Two basically divergent approaches to the turbulence problem appeared from the very beginning. O. Reynolds [4] and J.W. Strutt (Lord Rayleigh) [5] proceeded from the assumption that a laminar flow turned into the turbulent owing to its stability loss. In this case, the well-known method of small perturbations might be applied, and the problem could be reduced to finding out whether a small perturbation was damping and eventually faded away or was increasing in time. Then the primary exciting motion was automatically implied to be some periodic function of small amplitude being described by the same equations as the flow itself. Generally, the theory assumed the fluctuations of any form convenient for a given application, though a harmonic function was usually used.

H. Lorentz [6], on the contrary, allowed for any form of the primary exciting motion and looked for motion descriptions that were compatible only with the continuity equation. This approach to the study of turbulence did not develop further due to a lack of proper understanding of the primary perturbation image, at that time. Meanwhile the validity of Lorentz's idea can be easy demonstrated by the example of some constant turbulent flow, in which vectors of the main flow velocity and of the perturbation velocity are utterly different functions. Really, the flow velocity vectors of stationary flow do not depend on time, $\boldsymbol{U}(\boldsymbol{r})$, whereas fluctuations exist in the form of a spatio-temporal field $\boldsymbol{u}(\boldsymbol{r}, t)$ depending on time explicitly.

Absence of a physically proved form of the primary exciting motion, as well as the scientific authority of Rayleigh who explored the flow stability even before the turbulence phenomenon had been discovered, resulted in the first approach being generally recognized. Also, L. Prandtl's effort promoted recognition of the first approach to a great extent.

Having discovered fluid adhesion to the channel wall [7], Prandtl distinguished a thin layer within the stream adjoining the wall. This layer differed from the other (main) part of the flow by the considerable influence of viscous friction forces on the fluid's motion. Prandtl introduced into practice the notion of a "boundary layer" for the thin part of the flow, and a "flow core" where the friction influence was insignificant and negligible. After that it became possible to consider the flow core as a separate main flow, and to apply Bernoulli's equation to its description.

Also, Prandtl came out with a suggestion, according to which the pressure $P$ in the flow in each cross-section was nearly constant, i.e., the pressures in the boundary layer and in the main stream are approximately equal and determined by Bernoulli's equation. For instance, if the velocity vector $\boldsymbol{U}$ in a stationary two-dimensional flow is directed along the axis of the channel and the $z$-axis is normal to the walls, then it follows from this assumption, that

$$\left.\begin{aligned}(dP/dz) &= 0,\\ (dP/dx) &= \text{const.}\end{aligned}\right\} \quad (1.2)$$

Hence, using equations derived from Navier–Stokes and continuity equations by some essential further simplifying assumptions (see, for example [8]), Prandtl managed to describe the flow within the boundary layer. Stability of the flow was now just defined by stability of the laminar boundary layer. With all this going on, the primary exciting motion is considered as a flat harmonic wave propagating along the stream. As before, the incremental negative value implies that the exciting motion is fading, whereas the positive one means the perturbation is strengthening, and, accordingly, the laminar flow conversion into the turbulent follows.

The method of small perturbations allows us to estimate conditions of the laminar flow stability loss, but does not describe the perturbations in a channel. As a matter of fact, the friction tangential stress $\sigma$ is necessary to know in order to solve the problem. The stress in a laminar boundary layer is given by Newton's friction law according to which

$$\sigma = \eta\,(dU/dz). \quad (1.3)$$

Here the viscosity factor $\eta$ is determined by the momentum flux density across the flow, i.e., by the momentum transmission caused by impacts of molecules from neighboring layers. In gases, for example, the viscosity factor $\eta=\varrho\lambda\upsilon/3$ depends on clear physical parameters, such as the molecule free path $\lambda$ and molecule thermal velocity $\upsilon$. Therefore the flow description in the laminar case is developed as a purely deductive theory, starting from equations for the viscous flow.

Elaboration of the deductive theory for the flow turbulent mode was impossible in view of the regrettable fact that there was no conception of momentum transmission. Therefore the theory elaboration was confined by semi-empirical approximation (see, e.g. [8]). Then the structure of the formula (1.3) was taken as a model to formulate the turbulent friction law corresponding to averaged velocity $U$, while conception of the momentum transmission remained unknown. Then there were introduced seeming tangential stress (shearing stress) $\sigma_t$ as a function of the mean velocity $U$, instead of tangential stress $\sigma$, and the factor of turbulent exchange $A_t$ instead of the coefficient of dynamic viscosity $\eta$, to formulate the turbulent friction law.

However, the turbulent exchange factor $A_t$, as distinct from the viscosity coefficient $\eta$, depends on the velocity $U$ distribution in a channel cross-section, and is not a physical constant. Therefore, the velocity distribution was required as well to derive a friction law for solving the set of hydrodynamic equations. The dependence was constructed on the basis of the mixing path hypothesis advanced and later improved by Prandtl [7, 9]. Von Kármán's similarity hypothesis [10] yielded an analogous result.

The mixing path notion was formulated as a macroscopic analog of the molecule free path λ in kinetic theory. Length $L$ of the path is defined as the distance across the stream to the neighbor layer, where difference between velocity of the fluid "turbulent mass" (fluid turbulent cluster) and of the layer destination became equal to the mean velocity of the longitudinal fluctuation.

The definition does not seem to imply a clear physical picture of the process, since it makes sense provided that every such turbulent mass keeps on being a single whole, saving the velocity longitudinal component during the motion along the trajectory $L$. Besides, it is worth emphasizing that the turbulent mass may travel across a stream only under the action of some force which is the pressure gradient in the flow. Hence the idea of a pressure transversal gradient in a flow cross-section does not correspond to the supposition expressed by formula (1.2) according to which zero gradient has been assumed.

In view of obvious disadvantages and contradictions of available interpretation of the phenomenon, L. Richardson [11] explained the turbulent fluctuations by onset and interaction of vortices during cascade process of their disintegration. At some value of Reynolds' number Re = $UL/\nu > \mathrm{Re}_{cr}$ the flow loses its stability, and there appear vortices of typical size and parameters such as $L_1 < L$ and $u_1 < U$, $\mathrm{Re}_1 < \mathrm{Re}$. If number $\mathrm{Re}_1$ is large enough and still exceeds some critical value $\mathrm{Re}_c$, nascent vortices serve as sources for new vortices of the second generation.

The disintegration process lasts until Reynolds' number Re, at some cascade step $n$, amounts to critical value $\mathrm{Re}_C$. After that the vortices get quasi-stable and dissipate gradually under the influence of the viscosity. Superposition of the dispersed vortices at various stages of their development produces a spatio-temporal field of fluctuations in the flow.

Here we should note that this attractive concept, as can be seen at first sight, is open to one serious question. The vorticity can arise only within the flow area where exists the viscous friction and transversal momentum transfer, i.e., in the boundary layer. However the vorticity penetration through the border of the boundary layer into the flow potential part is impossible by virtue of Thomson's theorem ([3], §8). Therefore, the vorticity extension to the complete flow badly needs a special substantiation which in its turn needs some serious study.

All the troubles emerging in turbulence interpretation within the frames of various approaches seem to originate from the formal description of the momentum transfer mechanism, without taking into account actual details of the mechanism. As a result, the available turbulence models appear interiorly inconsistent, and mismatch real physical processes. Therefore, turbulent fluctuations, owing to their disorder and incomprehensibility, have been classed as random phenomena.

The rationality of this new approach can be easily demonstrated, when the fluctuations are expressed in terms of explicit functions of the fluid hydrodynamic equations. Substituting relations (1.1) into a Navier–Stokes' equation, and subsequent time averaging of the result, lead to Reynolds's equation (see, for example [12]). When the density ρ and kinematic viscosity ν are constant values, and $\overline{u_i} = 0$, the equation takes the form

$$\frac{\partial U_i}{\partial t} + U_j \frac{\partial U_i}{\partial x_j} = -\frac{1}{\varrho}\frac{\partial P}{\partial x_i} + \frac{\partial}{\partial x_j}\left(\nu \frac{\partial U_i}{\partial x_j} - \overline{u_i u_j}\right), \tag{1.4}$$

where the fluctuations are represented by pair correlation functions $\overline{u_i\, u_j}$.

This equation along with a continuity equation make up a set of five relations containing seven unknown functions, i.e., the equations for turbulent flow do not represent a closed system. Two additional relations are required to come to the sought closed set of equations, and to express dependence of the fluctuations on mean values of the flow parameters. Unknown pair correlations $\overline{u_i\, u_j}$ in equation (1.4) can be expressed through the averaging of flow parameters, though there appear the triple correlations $\overline{u_i u_j u_k}$ instead of $\overline{u_i u_j}$, which in its turn depends on quadruple correlations, and so ad infinitum.

It is necessary to cut short the perpetual chain of equations at some phase to obtain a consistent set of equations. The pair correlations are known to correspond to the tensor of turbulent (Reynolds') stresses $\varrho \overline{u_i u_j}$, while correlations of higher orders have no certain physical interpretation. Therefore the missing equations are usually deduced from some model approximation based on interpretation of experimental data or formulated from inductive reasoning.

## 1.2 The next approaches

Seeing that the turbulence theory was an imperfect one, J. Taylor [13] turned his attention to a statistical approach. Assuming the field of fluctuations being homogeneous and isotropic, J. Taylor tried to describe the fluctuations using such statistical characteristics as correlation function

$$F = \overline{u_1 u_2} \Big/ \sqrt{\overline{u_1^2}}\, \sqrt{\overline{u_2^2}}\,. \tag{1.5}$$

However, inasmuch as the isotropic turbulence had actually no uniform structure, A.N. Kolmogorov [14] put forward the concept of local isotropic turbulence, where one did not consider correlations of velocities, but of their differences within the length of *l*:

$$l = \int_0^{d/2} F\,dr. \tag{1.6}$$

A similar idea was simultaneously offered by A.M. Oboukhov [15].

Kolmogorov supposed that isotropism of this kind existed in any part of a turbulent flow, including those volumes where the parameter gradients were considerable and the turbulence Reynolds' number

$$\mathrm{Re} = \sqrt{\overline{u^2}}\, l/\nu \tag{1.7}$$

was large enough. Then the dimensional theory allowed one to obtain the spectrum of local isotropic turbulence power, known as Kolmogorov's "law of –5/3":

$$\varepsilon = Cqk^{-5/3}. \tag{1.8}$$

Here the fluctuation kinetic energy $\varepsilon$ depends on wave number $k \sim 1/l$, on the rate $q$ of the turbulent energy dissipation within a unit volume, and on quantity $C$, named as Kolmogorov's constant.

Kolmogorov constructed the concept using several underlying postulates. Scaling invariance is applicable provided that transformation of length $l$, velocity $U$ and ratio $P/\varrho$ occur simultaneously,

$$\left.\begin{array}{l} l \Rightarrow \alpha l, \\ U \Rightarrow U^h, \\ P/\varrho \Rightarrow \alpha^{2h} P/\varrho, \end{array}\right\} \tag{1.9}$$

while Navier–Stokes' equation with average parameters is considered. Here $h$ stands for arbitrary values of the power index. Further, there exists a permanent flow of energy from large-scale fluctuations to those of fine-scale. Then, this energy dissipates in a high-frequency spectral region. At last, energy flux $q_l$ at scale $l$ depends on other parameter magnitudes of the same scale.

The requirement of scale invariance is not observed in the area $l$ where the length is large, $l \sim l_1$, of the order of the flow size. This region includes a domain in which the fluctuations are generated. On the other hand, the invariance fails, if the length scale is too small, $l \sim l_2 \sim (\nu^3/q)^{1/4}$, because viscosity leads to an appreciable dissipation under this condition. Thus the scale invariance requirement is valid, and the value of $h$ can be estimated inside the so-called inertial interval $l_1 << l << l_1$.

However, the value of $h$ can be determined only under some additional hypotheses even within this interval. To get out of the situation, the turbulent energy dissipation rate, averaged over the interval, is supposed to be invariable. Then the dimensional analysis leads to the quantitative relation

$$q_l \sim u_l^3/l, \tag{1.10}$$

whence the Kolmogorov–Oboukhov law follows ([14, 15]):

$$u_l \sim (q_l\, l)^{1/3}\,. \tag{1.11}$$

A great number of works have been devoted to further development and generalization of the statistical description of turbulence (see, e.g. [16–21]). Naturally, steady flows were considered only in the frame of the statistical approach, and the problems of paramount importance, concerning turbulence origination and stochastization were not discussed at all.

D. Ruelle and F. Takens [22, 23] started exploration of these processes by means of chaotic dynamics which dealt with stochastic behavior of nonlinear deterministic processes. Chaotic dynamics employs the concept of a strange attractor which represents a multitude of attracting unstable trajectories in a phase space of the dissipative dynamic system under consideration. One of the main parameters characterizing a dynamic system is the Poincaré cross-section of the system.

The dynamic chaos idea as applied to turbulence and the strange attractor adaptability to a description of turbulent disorder origination are grounded on bifurcations in the stream, and opportunity of the strange attractor restoring itself after a transition to disorder. The restoration of an attractor has been successful with the flow of some special geometry, such as in the case of Taylor–Couette flow.

Some inconstancy of exponent $h$ is interpreted within the framework of this approach as a result of its quantity distribution over structures of different scales. And the turbulence itself is perceived as a spatio-temporal chaotic field of the structures interacting with each other. The chaotic flow dynamics is determined, much as in [11], by a small number of independent large-scale perturbations which then form coherent structures. Progress in application of chaotic dynamics to turbulence and to development of structures in the flows of various types is presented, for instance, in reports of the Symposium mentioned in [21].

Generally, the strange attractor concept relates to the matter of self-oscillatory systems, and hence turbulence as a phenomenon should be considered as a self-oscillatory process. However P.S. Landa [24] has cast serious doubts on the validity of such a consideration, at least with regard to subsonic flow in turbulent jets. Solutions of Navier–Stokes equations by the Krylov–Bogolyubov asymptotic method, and determination of the flow parameters in a jet [25] have strengthened the doubts.

According to the solution, the turbulence and coherent structures arise not due to self-excited oscillations, but by amplification of casual acoustic wave perturbations that exist always in a jet at the nozzle outlet. The velocity and pressure oscillations calculated coincide with data of many experimental researches. In particular, the results describe very well the velocity oscillation frequency shift towards a low-frequency region, as well as an oscillation space scale increase with the distance from the nozzle outlet.

As is clear from [25], acoustic waves even of weak intensity play a rather important role in formation of turbulent fluctuations. Dependence of turbulent flow characteristics in jets on acoustic waves has been investigated in many works, for example in [26]. Data of the measurements demonstrate clearly that acoustic oscillations, and, pay attention, the nozzle vibrations have similar considerable effect on turbulence properties. The effect depends on the oscillation frequency. At high frequency, turbulent fluctuations become weaker, whereas at low frequency they grew stronger. Results of these researches have been used for elaboration of a new method for flow parameter control [26–28].

One of the important findings of the researches [24, 27] consists in a profound analogy between changes in behavior of the turbulent jet and swing of a pendulum under influence of acoustic waves and casual displacement of a mechanical pendulum axis, accordingly. It should be noted in this connection, that the Navier–Stokes equation has nothing to do with the pendulum. Hence it is the ordinary wave equation that serves as a basis for the just mentioned analogy.

The turbulence research technique listed above has been creative as applied to different experimental conditions. Furthermore, a macroscopic description of the observable processes used in the technique does not need modeling of an exciting motion. Therefore there is not surprising that various methods have developed and appeared absolutely independently of one another. Nevertheless, there should exist some general physical basis, or pattern of an exciting motion inherent to the turbulence in all its manifestations. The pattern should explain origination of an observable spatio-temporal field of chaotic fluctuations, practically independent of initial perturbations and exterior noises.

L.D. Landau [29] and then E. Hopf [30] considered the primary exciting motion in the form of a harmonic function within the framework of stability theory. They analyzed behavior of the stream function for two-dimensional stationary flow. The fluid velocity along the wall in direction of the x-axis was supposed to depend only on the distance $z$ to the wall.

According to representation (1.1), some small harmonic exciting motion in the form of a one-dimensional wave was imposed on the stream function $\psi_0$ of the main unperturbed flow ([3], §41):

$$\psi = \varphi(z)\exp[i(kx - \omega t)]. \tag{1.12}$$

The stream function $\psi_0$ of cooperative motion $\psi_0+\psi$ satisfies the hydrodynamic equations by definition. As to stream function $\psi$ (1.12), its substitution into the linearized hydrodynamic equations gives an ordinary differential equation of the fourth order, an Orr–Zommerfeld equation (see, e.g. [8]):

$$(c-U)\left(\varphi''-k^2\varphi\right)+U''\varphi=\frac{i}{k\,\mathrm{Re}}\left(\varphi''''-2k^2\varphi''+k^4\varphi\right). \quad (1.13)$$

Equation (1.13) describes amplitude φ(z) of the stream function ψ, and underlies the stability theory of the laminar flow. It can be easily reduced to a dimensionless form. For this purpose all lengths are divided by the typical channel size *d*, and velocities are divided by the greatest velocity $U_0$ of the main flow, where Re = $U_0 d/\nu$ is Reynolds' number, *c* denotes velocity of the perturbation propagation. Its dimensional quantity equals $c = \omega/k$.

The equation's left-hand side includes inertial terms of the dynamical equation, while the right-hand side contains the terms relating to friction phenomena. Both components of perturbation velocity vanish on the wall surface. Therefore, boundary conditions at borders of the flow moving along the channel constituted by two planes (z = 0, *d*), or moving along a single wall (z = 0, ∞), should satisfy the requirements $\varphi = \varphi' = \varphi'' = 0$.

Study of laminar flow stability in relation to perturbations of type (1.12) is reduced to an eigenvalue problem for equation (1.13). It is necessary for the solution to specify basic flow *U*(*z*) and a Reynolds number Re. Varying wave number *k* for a pair of the values of Re and *k* allows one to determine an eigenfunction φ(*z*) and a complex frequency eigenvalue $\omega + i\gamma$.

If γ is a negative value, the perturbation damps in time. Otherwise the instability develops, and the perturbation exponentially increases. The intermediate case, γ = 0, corresponds to the so-called neutral curve *k*(Re), where Reynolds number Re amounts to critical value $\mathrm{Re}_{cr}$, i.e., to a value exceeding which means progressive intensification of the perturbation.

Landau and Lifshitz ([3], §26) estimated the perturbation wave development in time, remaining in the range of small perturbation method applicability. The harmonic perturbation velocity ***u*** was presented in the form of the product of two independent functions of coordinates and time,

$$u = \mathbf{A}(t)\,\boldsymbol{F}(\boldsymbol{r}). \quad (1.14)$$

The function $\boldsymbol{F}(\boldsymbol{r})$ relates to spatial distribution of the wave field, and A(*t*) corresponds to the velocity complex amplitude provided that $\gamma \ll \omega$:

$$\mathrm{A}(t)=\mathrm{A}_0\, e^{\gamma t} e^{-i(\omega t+\beta)}. \quad (1.15)$$

Here β in the imaginary exponent is some initial phase of the wave. When the Reynolds number satisfies requirement $\mathrm{Re} \geq \mathrm{Re}_{cr}$, there can exist oscillation frequencies in the perturbation spectrum, at least one frequency, at which the negative exponent γ reverses the sign to become positive, $\gamma > 0$.

Expansion of function A(*t*) into a power series enables one to estimate the time derivative from the oscillation amplitude squared, $|\mathrm{A}|^2 = \mathrm{AA}^*$. Averaging of the derivative within the time interval

$$2\pi/\omega \ll t \ll 2\pi/\gamma \tag{1.16}$$

leads to an equation for $|A|^2$. The equation to within the terms of the fourth order of $|A|$ looks like

$$\frac{d|A|^2}{dt} = 2\gamma|A|^2 - \alpha|A|^4. \tag{1.17}$$

Here the sign of averaging over the left- and right-hand items is omitted, since the averaging time is rather short according to equation (1.16). A factor α known as Landau's constant can be thought of as a positive value, when the flow instability is considered in relation to small initial perturbations.

Analysis shows that the flow instability originates from any arbitrarily small exciting motion, certainly provided $Re > Re_{cr}$. This equation for the exciting motion has the solution:

$$|A|^2 = \left(\frac{\alpha}{2\gamma} + \text{const} \cdot e^{-2\gamma t}\right)^{-1}, \tag{1.18}$$

which means that $|A|^2$ tends asymptotically to the following finite limit:

$$|A|^2_{max} = \frac{2\gamma}{\alpha}. \tag{1.19}$$

Generally, parameter γ is some function of Reynolds number Re. And inasmuch as $\gamma(Re_{cr}) = 0$ by definition, it can be expanded into a series. Using only the first item of the series, namely $\gamma \sim (Re - Re_{cr})$, and substituting it into equation (1.19), we can see that the amplitude $|A|$ depends on the square root of the difference $Re - Re_{cr}$ :

$$|A|_{max} \sim \sqrt{Re - Re_{cr}}\,. \tag{1.20}$$

This consideration allows us to draw some important conclusions [3, §26].

Absolute instability of the flow under the influence of perturbations like equation (1.12) results in periodic unsteady motion, provided $Re > Re_{cr}$. While the difference $Re - Re_{cr}$ is not too large, the motion can be represented as a sum of two distinct functions corresponding to the stationary motion and some additional periodical perturbation $\mathbf{u}(x,y,z,t)$ of small, but finite amplitude. This amplitude increases with Re as in equation (1.20). The velocity distribution in this flow is of the form

$$\mathbf{u} = \mathbf{f}(\boldsymbol{r})\exp(-i(\omega t + \beta)), \tag{1.21}$$

where $\mathbf{f}(\boldsymbol{r})$ is a complex function of the coordinates, and β denotes some initial phase.

For large $Re - Re_{cr}$, the separation of the velocity into two components is no longer meaningful. We then have simply some periodic flow with frequency ω. If, instead of the time, the phase $\varphi = \omega t + \beta$ is used as an independent variable, then the function $\mathbf{u}(\boldsymbol{r},\varphi)$ becomes a periodic function of the phased φ with period 2π. However, this function is no longer a simple trigonometric function, for Fourier series of this function,

$$\mathbf{u}=\sum_p \mathbf{A}_p(x,y,z)e^{-i\varphi_p}, \tag{1.22}$$

contains terms not only at the base frequency ω, but $p$-th harmonic frequencies $\omega_p$ as well.

Hydrodynamic equations for a turbulent stationary flow under fixed boundary conditions are known (see, e.g. [3, 8]) to have an unambiguous solution. This solution allows us to obtain the flow parameters averaged over time, and to compute distribution of the parameters. It should be emphasized that the distribution of the averaged parameters is completely determined by chosen external conditions, and it does not possess any degree of freedom.

As to the parameter fluctuations, this is quite another matter. The exciting motion in the form (1.12) gives merely an opportunity to estimate absolute value (1.19) of the complex amplitude of the oscillations. As a matter of fact, distribution of the oscillations, which even in steady-state flow wear a complicated appearance of subtle a spatio-temporal field, can not be understood. Then there is nothing to say about an opportunity to describe features and development of the fluctuation field. Also, unlike the averaged parameter, the initial phase of the harmonic perturbation wave turns out to be indefinite. Moreover, its physical interpretation is not quite clear.

Thus, the periodic exciting motion is not determined unambiguously by those external conditions that form the properties of the averaged flow parameters. Unlike the flow, fluctuations bear a degree of freedom caused by uncontrollable factors. In other words, the exciting motion form (1.12) does not reveal the fluctuation nature, and does not disclose a physical picture of the turbulence.

## 1.3 A new approach

In search of the exciting motion form, let us turn our attention to Reynolds' criterion. The physical sense of this criterion is known, and it can be easily demonstrated by an example of the laminar flow of a viscous incompressible fluid moving in the channel formed by two parallel planes at $z=d$ distance.

Let $\boldsymbol{J}$ be the vector of the momentum flux density of the fluid. Component $i$ of the vector in the direction of normal $\mathbf{n}$ to a surface element can be expressed through a tensor of the momentum flux density $\Pi_{ik}$ [3, §15]:

$$J_i = \Pi_{ik} n_k = p n_i + \varrho\, U_i U_k n_k - \sigma'_{ik} n_k \,. \tag{1.23}$$

The first item in the expression (1.23) is a force of pressure acting in the fluid. The second item relates to the reversible transference of the momentum connected with mechanical transfer of fluid elements. And the third item describes the viscous strain tensor,

$$\sigma'_{ik} = \eta\left(\partial U_i/\partial x_k + \partial U_k/\partial x_i\right) n_k, \tag{1.24}$$

determining the part of the momentum flux that is directed along the same momentum gradient. This part of the flux is not connected immediately with the fluid transference. It arises by virtue of the momentum flux nonreversible transference caused by the viscous friction and accompanying dissipation.

We consider a steady flow along axis $x$ of the channel ($V = W = 0$) with the velocity distribution:

$$U = U_0[1-(2z/d-1)^2]. \tag{1.25}$$

As it follows from equation (1.23), $x$- and z-components of the momentum flux in the channel middle, $z = d/2$, are accordingly

$$\left.\begin{aligned} J_x &= p + \varrho U_0^2, \\ J_z &= p. \end{aligned}\right\}. \tag{1.26}$$

The fluid stops moving at the wall surface, $z = 0$, and thus does not participate in the momentum transference. Hence we can write:

$$\left.\begin{aligned} J_{x0} &= p, \\ J_{z0} &= p - 4\eta\left(U_0/d\right). \end{aligned}\right\} \tag{1.27}$$

The second item of component $J_{z0}$ appears due to viscous friction which is responsible for some additional momentum flux in the surface normal **n** direction. This flux is of the density

$$j_{z0} = \sigma'_{xz} = \eta\left(\partial U/\partial z\right) = 4\eta\left(U_0/d\right). \tag{1.28}$$

Now, writing down the proportion, we will gain an expression for the Reynolds criterion accurate within a numerical factor:

$$\frac{J_x}{J_{z0}} = \frac{\varrho\, U_0\, d}{\eta} = \mathrm{Re}. \tag{1.29}$$

This expression means that the Reynolds criterion identifies the portion of the flow momentum that the fluid loses to overcome the friction force, i.e., its inverse value, to be exact. The same can be said with regard to the flow energy, for, as it is known, momentum transference always involves energy transfer.

The expressions (1.28) and (1.29) written for steady flow parameters averaged over time contains no information on the momentum transference mechanism. Note in this connection, that viscosity entails the flow eddying and the flow energy dissipation at the wall ([3], §79). The dissipation excites some local perturbations of temperature and pressure. It is obvious, that the temperature and pressure perturbations can not be accumulated anywhere within the steady flow, and thus they disperse over the volume.

The temperature perturbations are smoothed out by heat conductivity and by heat transfer to the wall. The pressure perturbations should propagate all over the flow as well, being converted into heat in the end. The process of the propagation is well known ([3], §70). A small volume of sudden pressure perturbation emits a spherical acoustic wave packet. It bears a momentum flux, thereof average density in the z-axis direction is

$$\overline{j_z} = \overline{\Pi_{zz}} = \varrho \overline{w^2} \, . \qquad (1.30)$$

The wave packet derives its strength from the main flow, or finally from the source supporting the whole fluid motion. Total momentum flux produced by many of such packets corresponds to the momentum flux defined by equation (1.28). To the point, each wave packet contains oscillating parameters ([3], §70). Superposition of the oscillations in the flow volume appears as a spatio-temporal field of the fluctuations.

This process is by no means in conflict with the Navier–Stokes equation which is replaced by two equations for velocity and pressure correspondingly. The first concerns the eddy motion, and the latter is an equation of the Poisson type ([3], §15). Its particular solution represents pressure distribution near the wall, which corresponds to the pressure perturbations mentioned. The general solution (Laplace's equation solution) determines harmonious waves similar to the emitted acoustic waves which follow from the wave equation. In its turn, the wave equation is deduced from Navier–Stokes and continuity equations after some simplifications.

The above propositions impel one to suggest that acoustic waves can play a more important role in formation of turbulent fluctuations than formerly believed, and that turbulent fluctuations can be of a wave nature. So that examination of a flow stability by means of superposition of the perturbations of the wave type on the main flow might not be only a matter of chance.

Note, in addition, a strong influence of outside sound strong influence on the turbulent properties of jets, as well as the analogy between turbulent fluctuations and pendulum oscillations, observed in [24, 27], serve as auxiliary arguments in favor of the wave nature of turbulence.

In the next chapters of the book we consider the “wave” concept of turbulent fluctuations. Fluctuations in the frame of this concept are interpreted as perturbations carried by acoustic waves. Some simple and obvious assertions underlie this consideration. In short they are reduced to the following theses [31–37]:

1. A viscous fluid (liquid or gas) moving in a channel loses some part of its energy to wall friction along the channel, which decelerates the flow. Any outside source, e.g., a piston, can serve to offset losses for flow maintenance.

2. The piston is able to transfer its energy over all of the flow by radiating a sequence of elementary pressure waves of small, though finite, amplitude. A sequence of waves forms a so-called simple wave, in which the fluid moves in the same direction as the piston.

3. Any channel has naturally a limited cross section. This means that the simple wave is actually the wave beam, and therefore wave fronts may not be flat but spherical by virtue of the law of Rayleigh’s diffraction divergence. Such waves are inevitably reflected by the walls.

4. Each acoustic wave encloses longitudinal oscillations of the hydrodynamic parameters. The amplitudes of oscillating velocity, pressure and temperature jump at the running spot of the wave reflection near the wall. If the wave is of enough strength and of appropriate angle of incidence, a local tangential discontinuity of the boundary layer develops in the vicinity of the spot. Instability of the tangential discontinuity is known to lead to a substantial local pressure increase ([3], §29).

5. The tangential discontinuity along with the following pressure rise develop while the spherical wave interacts with the boundary layer during local reflection. Since this process takes a very short time in comparison with other motions in the flow, it can be accepted as being momentary. Then the pressure rise zone propagates in the flow in the form of the acoustic wave packet. The packet thickness corresponds to the zone initial diameter, or its linear dimension.

6. The boundary layer within the simple wave is characterized by a strong velocity gradient and hence by vorticity. The current lines of the layer and the vortices do not intersect the surface that separates the area of vortex flow from potential motion, as Thomson’s (Kelvin) theorem runs ([3], §35). However the wave packet easily penetrates through the surface. For all that, the wave transfers the parameter oscillations into the main stream. At the same time, the convective component of the oscillating velocity carries over thereto some quantity of the fluid along with the frozen-in vorticity.

7. Superposition of a sequence of the wave packets results in a spatio-temporal field of flow parameter perturbations. Simultaneously the vorticity carried over by the packets creates its own system of perturbations, which is similar to a well-known interpretation [11]. This

implies two mechanisms of turbulent fluctuations. The difference between the two lies in their natures, which become apparent when one observes the effect of damping on the fluctuation rate. The fluctuations of the first kind are in essence oscillations, to call things by their proper names. In this case the wave amplitude fades away with the distance covered as $r^{-1}$, whereas the convective component of the oscillating velocity, responsible for vorticity transposition, is proportional to $r^{-2}$.

Oscillations of acoustic nature are observed not only in turbulent flows, but in turbulent flames and in detonation structures (see Chapter 9). Similar structures have been discovered in the plasma channels created by Besselian beams of laser radiation as well (see Chapter 10).

Coming back to the beginning of the chapter, remember an idea of H. Lorentz. He searched for such a form of the initial exciting motion, which would have some clear physical interpretation and proper equations differing from the equations for the flow itself. Now we might say that he seems to have been quite right in his quest for a new initial motion.

Hereinafter we will consider in detail the initial exciting motion, definition of the direct turbulence problem, properties and development of the oscillation spatio-temporal field, the process of the field stochastization, formulation of the inverse turbulence problem and an example of its solution.

To facilitate understanding of this new and unusual concept of turbulence, and to simplify mathematical expressions to a fewer number, the exploration is confined within fluids (liquid or gas) moving slowly along a wall or in channel

# 2
# Fluid motion

## 2.1 Equations of fluid motion

Slow motion of a fluid with equilibrium thermodynamic properties is characterized by velocity $\boldsymbol{U}(U, V, W)$, pressure $P$ and density $\varrho$. These symbols correspond to the parameter in equation (1.1), marked by the subscript f which is omitted here and hereinafter for writing simplification.

As the functions of the coordinates $x$, $y$, $z$ and time $t$, all these quantities are determined by the hydrodynamic equations formulating in differential form the laws of conservation of mass (continuity equation), of momentum (motion equation) and energy (energy equation). The greatest computational complexity in the set of hydrodynamic equations is bound up with the nonlinear motion equation. The most general form of the equation for an isotropic viscous fluid looks like ([3], § 15)

$$\varrho\left(\frac{\partial U_i}{\partial t}+U_k\frac{\partial U_i}{\partial x_k}\right)= -\frac{\partial P}{\partial x_i}+\frac{\partial}{\partial x_k}\left\{\eta\left(\frac{\partial U_i}{\partial x_k}+\frac{\partial U_k}{\partial x_i}-\frac{2}{3}\delta_{ik}\frac{\partial U_l}{\partial x_l}\right)\right\}+\frac{\partial}{\partial x_i}\left(\zeta\frac{\partial U_l}{\partial x_l}\right). \tag{2.1}$$

The quantity $\eta$ is the coefficient of viscosity, and $\zeta$ is called of second viscosity. Both of them are functions of pressure and temperature. Generally, parameters $P$ and $T$ vary within the flow, so that the viscosity coefficients $\eta$ and $\zeta$ change as well. However, the dependences of the coefficient quantities on the parameters $P$ and $T$ are rather weak in most slow flows. Therefore the quantities are customarily regarded as constants. In this case the equation becomes simpler, and it is possible to be presented in the vector form of a Navier–Stokes equation,

$$\varrho\left(\frac{\partial \boldsymbol{U}}{\partial t}+(\boldsymbol{U},\mathrm{grad})\boldsymbol{U}\right)=-\mathrm{grad}P+\left(\zeta+\frac{\eta}{3}\right)\mathrm{grad}\,\mathrm{div}\boldsymbol{U}+\eta\,\Delta\boldsymbol{U}. \tag{2.2}$$



L.N. Pyatnitsky, *Turbulence Nature and the Inverse Problem*, Fluid Mechanics and Its Applications 89, 17–27.

Then all the set of hydrodynamic equations are written in a vector form, as follows,

$$\left.\begin{aligned}
&\frac{\partial \varrho}{\partial t} + \operatorname{div} \varrho \boldsymbol{U} = 0, \\
&\varrho\left(\frac{\partial \boldsymbol{U}}{\partial t} + (\boldsymbol{U}, \operatorname{grad})\boldsymbol{U}\right) = -\operatorname{grad} P + \\
&\qquad + \left(\zeta + \frac{\eta}{3}\right) \operatorname{grad} \operatorname{div} \boldsymbol{U} + \eta \Delta \boldsymbol{U}, \\
&\frac{\partial}{\partial t}\left(\varrho\varepsilon + \frac{\varrho U^2}{2}\right) = -\operatorname{div}\left[\varrho \boldsymbol{U}\left(\varepsilon + \frac{U^2}{2}\right) + P\boldsymbol{U}\right] + \varrho q.
\end{aligned}\right\} \tag{2.3}$$

Here $\varepsilon$ is the internal energy per unit mass, and $q = q_+ - q_-$ stands for that specific energy which the fluid acquires from some external source to support its motion.

The reductive equations (2.3) present another complicated nonlinear set, and essential simplifications are needed to formulate the solution. The ideal fluid motion is one of such simplifications, when the viscosity, as well as the heat exchange between a fluid and the environment, and the fluid's different segments are considered as negligible, $q = 0$. Zero heat transfer means that the motion is adiabatic, i.e., the fluid entropy remains constant, $s = s_0 = \text{const}$.

Total energy of the fluid volume unit does change at these limitations merely owing to the energy transfer through this volume surface, and to the work of a pressure force over the fluid. Then the momentum conservation law is expressed by Euler's equation instead of Navier–Stokes' equation,

$$\frac{\partial \boldsymbol{U}}{\partial t} + (\boldsymbol{U}, \operatorname{grad})\boldsymbol{U} = -\frac{1}{\varrho} \operatorname{grad} P. \tag{2.4}$$

The thermodynamic relation at the adiabatic motion,

$$\operatorname{grad} H = (1/\varrho) \operatorname{grad} P,$$

including enthalpy $H = \varepsilon + P/\varrho$, can be employed in equation (2.4) which can be written in the form

$$\frac{\partial \boldsymbol{U}}{\partial t} + (\boldsymbol{U}, \operatorname{grad})\boldsymbol{U} = -\operatorname{grad} H. \tag{2.5}$$

The fluid motion being adiabatic does not modify the continuity equation, and the equation set (2.3) reduces to the set

$$\left.\begin{aligned} &\frac{\partial \varrho}{\partial t} + \operatorname{div} \varrho \boldsymbol{U} = 0, \\ &\frac{\partial \boldsymbol{U}}{\partial t} + (\boldsymbol{U}, \operatorname{grad})\boldsymbol{U} = -\operatorname{grad} H. \end{aligned}\right\} \tag{2.6}$$

At the same time, using a formula

$$1/2 \operatorname{grad} U^2 = [\boldsymbol{U}, \operatorname{curl}\boldsymbol{U}] - (\boldsymbol{U}, \operatorname{grad})\boldsymbol{U}$$

which is well-known in vector analysis, Euler's equation can be rewritten in the form

$$\frac{\partial \boldsymbol{U}}{\partial t} - [\boldsymbol{U}, \operatorname{curl}\boldsymbol{U}] = -\operatorname{grad}\left(H + U^2/2\right). \tag{2.7}$$

Now consider steady flow, when $\partial \boldsymbol{U}/\partial t = 0$, and define the concept of the stream line. It is a line such that the tangent to it at any point indicates the direction of the fluid velocity at that point. The line is determined by the following differential equations:

$$\frac{dx}{U} = \frac{dy}{V} = \frac{dz}{W}. \tag{2.8}$$

In a steady flow the streamlines do not vary with time, and they coincide with the paths of the fluid particles. This coincidence does not occur in non-steady flow.

The vector $[\boldsymbol{U}, \operatorname{curl}\boldsymbol{U}]$ is perpendicular to $\boldsymbol{U}$, and its projection on the direction of $\boldsymbol{l} = \boldsymbol{U}/U$ is equal to zero by definition, whereas the gradient projection onto the same direction is the derivative in that direction. The projection of the gradient on the direction $\boldsymbol{l}$ is the derivative in that direction. Then equation (2.7) reduces to

$$\frac{\partial}{\partial l}\left(H + U^2/2\right) = 0.$$

Hence, along the line (2.8) is fulfilled the relation known as Bernoulli's equation,

$$H + U^2/2 = H_0 + U_0^2/2 = constant. \tag{2.9}$$

It is worth underscoring here the distinction between steady and non-steady flows. In a steady flow the stream lines do not vary with time, and they coincide with the paths of the fluid particles. In a non-steady flow this coincidence does not occur, and the tangents to the stream lines indicate the directions of the fluid particle velocities in the space at various points at a chosen instant. The tangent to the path indicates the direction of the velocity of a chosen fluid particle at various times.

Note in addition, an important difference between Bernoulli's equation for potential flow and that for general case of flows. In the general case, the *constant* in equation (2.9) does not change along any chosen stream line, but different stream lines have various values of the *constant*. In potential flow it keeps the same quantity throughout all the fluid.

## 2.2 Vorticity

An important parameter of fluid flow is velocity circulation $\boldsymbol{\Gamma}(\boldsymbol{U})$ of the fluid's particles around some closed contour $L$ drawn in the fluid at a given instant,

$$\Gamma(\boldsymbol{U}) = \oint_L \boldsymbol{U} d\boldsymbol{l}. \tag{2.10}$$

The contour moves along with those fluid particles in the flow. The velocity circulation evolution in time is determined by the total time derivative on the moving contour ([3], § 8),

$$\frac{d}{dt}\oint_L \boldsymbol{U} d\boldsymbol{l} = \oint_L \frac{d\boldsymbol{U}}{dt} d\boldsymbol{l} + \oint_L \boldsymbol{U} \frac{d(d\boldsymbol{l})}{dt} d\boldsymbol{l}. \tag{2.11}$$

The second integral in the sum of the equation vanishes as it is the integral from the total differential around the closed contour, and remains valid,

$$\frac{d}{dt}\oint_L \boldsymbol{U} d\boldsymbol{l} = \oint_L \frac{d\boldsymbol{U}}{dt} d\boldsymbol{l}. \tag{2.12}$$

The derivative of the integrand on the right-hand side is known from Euler's equation, $d\boldsymbol{U}/dt = -\mathrm{grad}H$. Substituting the derivative $d\boldsymbol{U}/dt$ into equation (2.12), replacing the integral along the contour by the integral over surface $S$ circumscribed by the contour $L$ according to Stokes' formula, and taking into account equality $\mathrm{curl}\,\mathrm{grad}H = 0$, we obtain,

$$\oint_L \frac{d\boldsymbol{U}}{dt} d\boldsymbol{l} = \int_S \mathrm{curl} \frac{d\boldsymbol{U}}{dt} d\boldsymbol{S} = 0. \tag{2.13}$$

Whence follows the final result

$$\frac{d}{dt}\oint_L \boldsymbol{U} d\boldsymbol{l} = \frac{d\Gamma}{dt} = 0, \text{ or } \oint_L \boldsymbol{U} d\boldsymbol{l} = \text{const} \tag{2.14}$$

Thus the velocity circulation around a closed contour within a moving fluid is constant in time. This is Thomson's (Kelvin) dynamical theorem, or the law of conservation of velocity circulation. According to Stokes' formula, the velocity circulation (1.10) can be expressed in terms of the vorticity $\boldsymbol{\Omega} = \mathrm{curl}\boldsymbol{U}$ at the fluid's given point,

$$\Gamma = \int_S \mathbf{\Omega}\, d\mathbf{S} = const. \tag{2.15}$$

When the area $S$ is getting rather small, $S \to 0$, equality (1.15) reduces to

$$\Gamma \approx \mathbf{\Omega}\, d\mathbf{S} = const. \tag{2.16}$$

The *const* value being nonzero means that the flow is vortical, and the product $\mathbf{\Omega} d\mathbf{S}$ constancy can be understood in the sense that the vorticity moves along with the fluid. If the vorticity is zero at some point of the flow, the condition $\mathbf{\Omega} = 0$ inheres in the whole stream line. When the condition const $= 0$ is fulfilled for all the stream lines of the flow, we deal with the fluid's potential motion.

Now, we consider the flow of viscous fluid in the approximation of incompressibility, when the density $\varrho$ variation is a negligible quantity, $\rho = 0$. The equations of continuity and Navier–Stokes from the set (2.3) in this approximation are simply written as

$$\left.\begin{aligned} &\operatorname{div}\boldsymbol{U} = 0, \\ &\frac{\partial \boldsymbol{U}}{\partial t} + (\boldsymbol{U}\,\mathrm{grad})\boldsymbol{U} = -\frac{1}{\varrho}\,\mathrm{grad}P + \nu\Delta\boldsymbol{U}, \end{aligned}\right\} \tag{2.17}$$

Here ν denotes the kinematic viscosity factor. Boundary conditions in a viscous fluid are determined by the adhesion phenomenon. According to the phenomenon, all the velocity components $\boldsymbol{U}(U, V, W, t)$ vanish on the wall surface,

$$U(t)\big|_{z=0} = V(t)\big|_{z=0} = W(t)\big|_{z=0} = 0. \tag{2.18}$$

The first equation in (2.17) does not depend on the fluid pressure. The second equation does, but it can be eliminated by means of the operation curl applied to both sides of the equation. Acting in this way and using some vector analysis formulas results in the new equation for $\mathbf{\Omega} = \mathrm{curl}\boldsymbol{U}$,

$$\frac{\partial}{\partial t}\mathbf{\Omega} = \mathrm{curl}[\boldsymbol{U}\mathbf{\Omega}] + \nu\Delta\mathbf{\Omega}, \tag{2.19}$$

where

$$\Delta\mathbf{\Omega} = \Delta\left(\frac{\partial W}{\partial y} - \frac{\partial V}{\partial z}\right)\mathbf{i} + \Delta\left(\frac{\partial U}{\partial z} - \frac{\partial W}{\partial x}\right)\mathbf{j} + \Delta\left(\frac{\partial V}{\partial x} - \frac{\partial U}{\partial y}\right)\mathbf{k}.$$

Since the fluid is incompressible, equation (2.19) can be transformed by expanding the product in the first term on the right side and applying the equation $\operatorname{div}\boldsymbol{U} = 0$ to the form ([3], §15):

$$\frac{\partial}{\partial t}\mathrm{curl}\boldsymbol{U} + (\boldsymbol{U}\nabla)\mathrm{curl}\boldsymbol{U} - (\mathrm{curl}\boldsymbol{U}\cdot\nabla)\boldsymbol{U} = \nu\Delta\mathrm{curl}\boldsymbol{U}. \tag{2.20}$$

There is only one unknown function in equation (2.20), and in (2.19) as well, namely velocity $\boldsymbol{U}$. If the velocity distribution is known, the pressure distribution in the flow can be in principle found by solving the equation of Poisson's type

$$\Delta P = -\varrho \frac{\partial U_i}{\partial x_k} \frac{\partial U_k}{\partial x_i} = f(x, y, z). \tag{2.21}$$

This equation is obtained by taking the operation div for both sides of Navier–Stokes' equation from the set (2.17). The function $f(x,y,z)$ specified in the area G has a partial solution at some sampling point $x, y, z$, given by the formula

$$P(x,y,z) = \frac{\varrho}{4\pi} \iiint_G \frac{f(\xi,\eta,\zeta)\, d\xi\, d\eta\, d\zeta}{\sqrt{(x-\xi)^2 + (y-\eta)^2 + (z-\zeta)^2}}, \tag{2.22}$$

where $\xi$, $\eta$, $\zeta$ are variables of the integration.

Equations (2.19) and (2.21) enable us to cast some light on the fluid motion's general properties and structure. The fluid viscosity in the flow kernel does not influence the hydrodynamic parameters and their distributions, for it is relatively small. So that factor $\nu$ in equation (2.19) may be neglected, and there is good reason to suppose $\nu = 0$. Then it follows from equation (2.19) that $\boldsymbol{\Omega} = 0$, and the fluid motion in the main part of the flow does not differ from the flow of an ideal fluid. The function $f(x, y, z)$ can be easily found as well.

On the contrary, the fluid viscosity in the boundary layer is relatively large, $\nu \neq 0$, and it causes a drag force which slows down the fluid motion. Consequently the vorticity $\boldsymbol{\Omega}$ within the layer may not be neglected, $\boldsymbol{\Omega} \neq 0$, and solution of equations (2.19) and (2.21) with $f(x, y, z) \neq 0$ comes as an intricate problem.

Solution of the boundary problem for Poisson's equation consists of two parts, $P \Rightarrow P + \mathcal{P}$, the partial solution $P(x, y, z)$ and general solution $\mathcal{P}$ which can be a constant ([12], §1.1). The first part presented by formula (2.22) depends on prescribed function $f(x, y, z)$. The second part is determined by Laplace's equation, $\Delta\mathcal{P} = 0$. The boundary problem solution for Laplace's equation contains harmonic functions of coordinates. Each of them is unambiguously selected by boundary conditions, i.e., by the pressure distribution on the border of region G of the function definition. Thus the friction force action in the flow boundary layer is accompanied by formation of the vorticity $\boldsymbol{\Omega}$ and of the pressure field according to functions $P$ and $\mathcal{P}$.

All modifications of parameters in a stationary flow are in a dynamic equilibrium, and any perturbation comes again to the equilibrium state during some transitional period $\tau$. This period depends on the sound (acoustic) velocity $c$, which is responsible for the perturbation spread.

If the period is short in comparison with duration $T$ of the parameter changes in the process under investigation, $\tau \ll T$, then the transitional process occurs practically without any imbalance in the flow. Otherwise the effect of the retardation is observed in the transient period, well-known and studied in electrodynamics as the delayed potential.

While the fluid is considered as incompressible, $\varrho \gg \rho \to 0$, the sound velocity $c = (\partial p/\partial \rho)_S$ increases infinitely, and rapid propagation of the perturbation wave allows us to neglect the retardation phenomenon.

## 2.3 Wave equation and incompressibility conditions

Actually the acoustic wave velocity in any fluid is a finite quantity. The wave implies an adiabatic oscillating motion of the fluid at small amplitude, which is alternately compression and rarefaction. Moreover, the pressure $p$ and density $\rho$ perturbations in the wave are bound by the linear relation that follows,

$$p = \left(\frac{\partial p}{\partial \rho}\right)_S = c^2 \rho. \tag{2.23}$$

The relation (2.23) replaces the energy equation in (2.3) in this case. While the averaged values of the parameters $\boldsymbol{U}$, $P$, $\varrho$ from the relations (1.1), satisfy the equation set (2.3), there remain in the set the small perturbations $\boldsymbol{u}$, $p$, $\rho$. As to quadratic terms of the perturbations, they can be neglected in the equations. Then the set of equations (2.3) is reduced to the following linearized equations ([3], §64):

$$\left.\begin{aligned} &\frac{\partial \rho}{\partial t} + \varrho \operatorname{div} \boldsymbol{u} = 0; \\ &\frac{\partial \boldsymbol{u}}{\partial t} + \frac{1}{\varrho} \operatorname{grad} p = 0. \end{aligned}\right\} \tag{2.24}$$

The operation curl applied to the second equation of the set produces the following equation [38]:

$$\frac{\partial \boldsymbol{\Omega}}{\partial t} = 0. \tag{2.25}$$

So, the acoustic perturbations really create a potential field. Then the field parameters can be expressed through the velocity potential $\varphi$ according to the relation $\boldsymbol{u} = \operatorname{grad}\varphi$. Such expressions can be easily derived from the second equation of the set (2.24) and formula (2.23):

$$p = -\varrho \frac{\partial \varphi}{\partial t}, \quad \rho = -\frac{\varrho}{c^2}\frac{\partial \varphi}{\partial t}, \quad \boldsymbol{u} = \frac{\partial \varphi}{\partial \boldsymbol{r}}. \tag{2.26}$$

Combining the first equation of the set (2.24) with the relations (2.26) results in the so-called the wave equation

$$\frac{\partial^2 \varphi}{\partial t^2} - c^2 \Delta \varphi = 0. \tag{2.27}$$

An elementary solution of the equation is expressed in terms of delta function $\delta(r-ct)$:

$$\varphi = \frac{\delta(r - ct)}{r}. \tag{2.28}$$

The solution features the perturbation, created by an instantaneous point source, in the form of a sphere $(r-ct)$ moving away from the radiant with the sound velocity and gradually decreasing amplitude.

Superposition of the elementary solutions gives an opportunity to analyze the perturbations from a more complicated radiant, such as the sources of the kind $f(R,c\tau)$. If the source radiates in a brief time, $c\tau \ll R$, the perturbation can be considered as instantaneous parameter distribution $f(R)|_V$ within the small volume V at the time instant $t = 0$.

To determine the response in some point $r$ to the perturbation of type $f(R)|_V$, the Green function $G(t,r)$ should be constructed for the given wave equation,

$$G(t,r) = \frac{1}{4\pi t}\, \delta(r - ct). \tag{2.29}$$

The problem concerning the space propagation of $f(R)$-type perturbation and, finally, related to the parameter spatio-temporal field created by the perturbation can be solved by convolution of two functions, $G(t,R)$ and $f(R)$, within all the explored volume V:

$$\begin{aligned} \varphi(t,r) &= \int_V G\big(t, |\boldsymbol{r} - \boldsymbol{R}| - ct\big)\, f(R)\; d\boldsymbol{R} = \\ &= \frac{1}{4\pi t} \int_V \delta\big(|\boldsymbol{r} - \boldsymbol{R}| - ct\big)\, f(R)\; d\boldsymbol{R}. \end{aligned} \tag{2.30}$$

Function $f(R)$ wants the explicit initial definition to evaluate the field $\varphi(t,r)$. Also it is necessary in presetting distributions of the functions in space and time to write down the boundary conditions for determination of the spatio-temporal field of the parameter perturbations. As a matter of fact, the question concerning the turbulence essence is reduced to a problem which implies knowledge of the primary perturbation nature and process of the perturbation formation. But all we know is that the velocity $\boldsymbol{u}$ should be equal to zero on a wall surface of a channel.

The price to be paid for transition to an image of an ideal fluid has turned out to be the fluid viscosity and the energy dissipation concerned with it. However both of these qualities can have considerable impact on the system behavior, and they should be taken into account in many tasks. In such cases, it is usual to neglect some other components of the equation set (2.3), to which the solution is the least sensitive in terms of the analytical result being closest to the real process under consideration.

It follows from Navier–Stokes' equation in (2.3), that the density changes make the least impact on the flow feature. Therefore one of the ways to simplify the equations consists in disregarding the compressibility, which means that assumption $\rho \ll \varrho$ $(\rho \to 0)$ may be employed instead of the representation $\varrho_f = \varrho + \rho$. The density perturbation $\rho$ depends on the flow velocity $U$, or on the value of Mach number $M = U/c$. According to ([3], §10), the density perturbation permissible quantity is estimated as being of the order of $\rho \sim \varrho M^2$. Therefore, the first condition of a fluid's incompressibility is the requirement of the flow being slow,

$$M \ll 1. \tag{2.31}$$

However, the requirement (2.31) is necessary and sufficient only for the steady flow. In a non-steady case the fluid incompressibility needs one more requirement, which limits the change rate of the local flow parameters. An estimation of the components in equations (2.3) shows that the density variation can be considered as small, if the typical variation time $\tau$ is great as compared with duration $T$ of the sound wave propagation along characteristic linear dimension $L$ of the explored volume ([3], §10):

$$T \gg L/c. \tag{2.32}$$

In other words, this additional requirement means that the effect of the retarded potential of the acoustic wave field gives a wrong description of the flow, if the inequality (2.32) is not obeyed. The retarded potential notion relates to the inhomogeneous wave equation, and is fully considered in ([39], §62). Certainly, the effect loses its force, if the length $L$ happens to be small, or if the signal (the sound wave in fluid) fades away too rapidly.

The sound wave amplitude fades in time as the exponential function $\exp(-\kappa ct)$, where $\kappa$ is the attenuation factor. Let number $N$ be the ratio of the wave amplitude at the start time to the amplitude in a later instant $T$. The interval $T$ can be named as the wave life time in the sense of, and equal to,

$$T = \ln N/(\kappa c). \tag{2.33}$$

For example, if $N = 100$ and pertinent frequency 100 кHz, then the perturbation life span will be $T \approx 0.03$ s in air ($\kappa \sim 4 \times 10^{-3}$ cm$^{-1}$) and $T \approx 0.8$ s in water ($\kappa \sim 40.37 \times 10^{-4}$ cm$^{-1}$),

The path length $L$ including reflections, which the perturbation wave covers in fact within an explored channel, can happen to be less than the distance that the wave runs during its life time, and $L/c < \ln N/(\kappa c)$

Hence the second requirement for the fluid being accepted as incompressible is written in the form

$$\tau \gg \frac{L}{c}, \frac{\ln N}{\kappa c}. \tag{2.34}$$

Thus the equation set (2.17) is valid, provided that the requirements (2.31) and (2.32), or (2.34), are satisfied. Hence, when the fluid density is accepted to be constant, $\rho = 0$, the set (2.17) is true automatically.

The parameters $\boldsymbol{U}$, $P$ and $\varrho$ mentioned in the equalities (1.1) are the time averaged values. The averaging procedure can be understood as averaging of equations (2.3) in time $t$ within the interval $T < t \ll \tau$. This procedure is similar to averaging of the perturbation (1.15) in the interval (1.16), that leads to equation (1.17). However, the assumptions employed in derivation of equation set (2.17) are not applicable to perturbations $\boldsymbol{u}$, $p$, $\rho$ introduced in equation (1.1), for typical duration $\tau$ of the perturbation local alternation is extremely rapid as compared with attenuation time $T$ of the wave.

In other words, the small perturbations $\boldsymbol{u}$, $p$, $\rho$ obey entirely the requirement (2.31) written now in the form $M = u/c \ll 1$. At the same time, the requirement (2.32) may not be satisfied in principle. Indeed, time $\tau$ of the perturbation wave change can be estimated as $\tau \sim \lambda/c$, whereas $T = L/c$. An obvious result of the comparison implies the inequality $\tau \lll T$ which conflicts with requirement (2.32). Therefore the fluid compressibility, i.e. condition $\rho \neq 0$, should be taken into account in the initial equations (2.3), while dealing with propagation of the perturbation waves.

Thus, it follows from all the statements above, that solutions of equations (2.3) may describe simultaneously several modes of fluid motion. These modes are the translational and eddy motions, as well as the potential vortex-free motion induced by acoustic waves. They coexist in the flow, though thereof spatio-temporal properties are essentially different.

For instance, the main flow potential motion and vorticity can be two- or three-dimensional, whereas the perturbation wave propagates always in all directions, and is certainly three-dimensional. The first two modes develop rather slowly at moderate motion of the fluid flow, and a stationary state of the parameters is possible. Contrariwise, the perturbation waves create some ephemeral field of the parameters, which remains changeable even in the flow's steady state.

Note, there can appear some secondary terms of motion–motion interaction besides the mentioned processes. However the analysis ([12], §1.7) shows that they are small, and consequently are not considered here.

The equation set (2.3) fits the description of all kinds of fluid motion, though it is hard to extract a general solution. But in the case under consideration, when the fluid moves slowly, and the perturbation wave runs rapidly, the set can be solved. The solution procedure becomes a transformation of the total equation system (2.3) into more suitable separate sets of equations for each kind of motion. So, equations (2.17) relate to incompressible fluid motion, and equations (2.24) concern propagation of perturbation waves. Remember, however, that the flow perturbations in the form of vorticity and wave motion have a single common source which is viscous friction.

To further clarify the physical picture of the viscous friction effect on flow qualities, the motion of the viscous fluid should be compared with the ideal fluid motion. The ideal fluid is known to slide on a wall surface without any friction, so that a function of the velocity tangential component $U$ undergoes a discontinuity at the surface. The discontinuity in the tangential velocity component is equivalent to an infinitely thin boundary layer. Therefore, some vorticity and perturbation of a finite amplitude can not arise in the ideal flow moving along the wall, if the surface is plane and there are no external sources of perturbations. The perturbations may arise due to the wall bend surface or under the action of some external sources.

Energy dissipation in a viscous fluid perturbs the flow parameters, and the momentum flux transferred by the viscosity diffuses the velocity jump at the wall. As a result, the boundary layer becomes of a finite thickness, and the velocity therein decreases gradually in the direction of the wall. At the same time, vortices appear, as equation (2.19) suggests, and the pressure perturbations appear, as equation (2.21) implies. For all that, the inequalities $M \ll 1$ and $t \ll T \ll \tau$ mean, that the field of the wave perturbations in a moderate flow comes to equilibrium state much faster than the averaged parameters do.

The perturbation, as it will be shown below, propagates as a wave of spherical shape. Such a wave intensity is inversely lproportional to the distance run squared, $r^{-2}$, and the perturbation dies away before the viscosity effect reaches a significant extent. Therefore, the perturbations propagate as isentropic waves, and formation of the perturbation spatio-temporal field may be considered as if the fluid were motionless. Then the parameters of the steady flow are expressed through the following formulas:

$$\left.\begin{aligned} \boldsymbol{U}_{\mathrm{f}}(\boldsymbol{r},t) &= \boldsymbol{U}(\boldsymbol{r})+\boldsymbol{u}(\boldsymbol{r},t), \\ P_{\mathrm{f}}(\boldsymbol{r},t) &= P(\boldsymbol{r})+p(\boldsymbol{r},t), \\ \varrho_{\mathrm{f}}(\boldsymbol{r},t) &= \varrho(\boldsymbol{r})+\rho(\boldsymbol{r},t). \end{aligned}\right\} \qquad (2.35)$$

Generally speaking, the functions $\boldsymbol{U}(\boldsymbol{r})$, $P(\boldsymbol{r})$, $\varrho(\boldsymbol{r})$ may depend on time as well. However, the analytical solution can, in practice, be completed only for an incompressible fluid in this case, i.e., when requirements (2.31) and (2.32), or (2.34) are satisfied.

# 3
# Distribution of parameters in viscous flow

## 3.1 Velocity profiles in a flow cross-section

Without losing the generality of our reasoning, and at the same time to be definite, we shall consider the simple case of a steady flow, when the fluid moves slowly ($M \ll 1$, $\rho = 0$) along the $x$-axis of the channel formed by two parallel plane walls located at $z = 0$ and $z = d$.

When we deal with an ideal fluid, Bernoulli equation (2.9) takes the simple form

$$\frac{U^2}{2} + \frac{P}{\varrho} = \text{constant}, \tag{3.1}$$

where the ratio $P/\varrho$ replaces the enthalpy $H$. The potential flow of incompressible fluid is described in a particularly simple manner that includes Laplace's equation for the velocity potential $\varphi$,

$$\Delta\varphi = 0. \tag{3.2}$$

The boundary conditions have to be added to equation (3.2). At the immovable wall surfaces, the fluid velocity component normal to the surface has to be zero, $U_n = \partial\varphi/\partial n = 0$. This solves the problem of ideal fluid flows.

As to viscous fluid motion, the parameter distributions are under the influence of fluid friction, including friction force at the channel walls. The fluid velocity is determined by the set of equations (2.17). In this case, the velocity $\boldsymbol{U}$ has components $U \neq 0$, $V = 0$, $W = 0$, and $U$ is a function of only ordinate $z$, $U(z)$. Then the equations (2.17) for a steady flow at the accepted conditions are written in the following simple form ([3], §17),

$$\left.\begin{aligned} &\frac{\partial^2 U}{dz^2} = \frac{1}{\eta}\frac{\partial P}{\partial x}, \\ &\frac{\partial P}{\partial z} = 0. \end{aligned}\right\} \tag{3.3}$$



L.N. Pyatnitsky, *Turbulence Nature and the Inverse Problem*, Fluid Mechanics and Its Applications 89, 29–42.

The second equation in (3.3) shows that the pressure is independent of the ordinate $z$. While the fluid pressure is constant in the $z$ direction, the right-hand side of the first equation should be a function of $x$ only, whereas the left-hand side is a function of $z$. This means that both sides of the equation are constant, and consequently

$$\frac{dP}{dx} = \text{constant}. \tag{3.4}$$

Thus the pressure is a linear function of the coordinate $x$ along the direction of the flow motion. Then we obtain the parabolic law of velocity distribution in a flow cross-section

$$U(z) = -\frac{1}{2\eta}\frac{dP}{dx}z^2 + a\,z + b. \tag{3.5}$$

The constants $a$ and $b$ are determined from the boundary conditions which imply that on the channel walls, $z = 0$ and $z = d$, the fluid is immovable, $\boldsymbol{U} = 0$. Then

$$U(z) = \frac{1}{2\eta}\frac{dP}{dx}z\,(d - z). \tag{3.6}$$

With the dimensionless values, when a point altitude $z$ is measured in scale of $d$, and the velocity maximum $U_0$ is located at the middle of the channel, $z = d/2$, relation (3.6) is written as

$$\frac{U(z)}{U_0} = 4\,z\,(1 - z). \tag{3.7}$$

A similar result can be obtained in the channels of other profiles. For instance, the velocity distribution in a pipe with circular cross-section by radius $R$ ($r \to r/R$) is

$$U(r) = (1 - r^2). \tag{3.8}$$

The distribution (3.7) takes its profile under the influence of the shearing stress, or friction force. The same force directed to the opposite side along the $x$-axis acts on surfaces of the walls. The force per the surface unit area is defined by the expression:

$$\sigma_{xz} = \eta\left.\frac{\partial U(z)}{\partial z}\right|_{z=0} = -\frac{d}{2}\frac{dP}{dx}. \tag{3.9}$$

This force equation represents the momentum $x$-component being transmitted by the fluid to the surface per unit time. There exists constant flux of the momentum $x$-component, which is responsible for the velocity gradient $\partial U/\partial z$ in the flow.

Formulas (3.5, 3.9) concern laminar flows. Similar momentum flux has to exist in the case of turbulent flows as well, though attributed to the gradient of the averaged velocity $U$. As follows from dimensional arguments and from measurements, it is

$$\sigma_{xz}=\varrho\left(\kappa z\frac{dU(z)}{dz}\right)^2, \tag{3.10}$$

where $\kappa$ is a numerical value known as a von Kármán constant that amounts to 0.4.

The gradient of the velocity $x$-component in the steady turbulent flow of a viscous fluid moving along a wall can be represented in an explicit form. Then the integration of equation (3.10) gives the logarithmical law for the velocity component distribution to within the integration constant C ([3], §42):

$$U=\sqrt{\sigma_{xz}/\varrho}\,(\ln z+\mathrm{C}). \tag{3.11}$$

However, the constant C can not be determined from boundary conditions in the usual way, for the solution (3.11) is defined neither near the wall nor far from it:

$$\left.\begin{array}{l} z\to 0,\ U\to\infty, \\ z\to\infty,\ U\to\infty. \end{array}\right\} \tag{3.12}$$

In view of the uncertainty of the relations, estimation of the constant C needs some additional reasoning. In the meanwhile, the information we seek lies in the fact that the dependence (3.11) leaves out of account the molecular friction at the wall. The molecular friction is important just in the immediate vicinity of the wall surface, within some rather thin layer $z < z_0$. The molecular friction in the layer is represented by the well-known relation $\sigma = \varrho\nu(dU/dz)$. Integration of this relation at $U|_{z=0} = 0$ results in the following simple velocity linear dependence on the distance $z$ from the wall,

$$U=\frac{\sigma}{\varrho\nu}z. \tag{3.13}$$

So then, the whole velocity profile can be invented as a composite dependence consisting of two parts joined end-to-end at the point $z_0$. One of the parts is the linear function (3.13) which is valid within the layer $z \le z_0$. As for the other part, a function similar to the logarithmic function (3.11) can be employed in the region of $z > z_0$. And the point $z_0$ can serve as that boundary condition which can help to estimate the constant C in the function (3.11). Following the appropriate mathematical routine, we can obtain the law (3.11) in the form

$$U = \sqrt{\sigma/\varrho}\ \ln\frac{z}{z_0}. \tag{3.14}$$

The form of the law evidently retains indeterminacy at the wide break in the region of $z \to \infty$. In the channel configuration we have chosen, this indeterminacy is responsible for the break of the velocity profile in the middle of the channel. On the other hand, there are abundant physical measurements in channels (see, e.g. [40]), which show the velocity $U$ profile being power-behaved with good accuracy. In particular, the flow velocity distribution between the channel's two walls is written as a universal function of the dimensionless argument $z \to z/d$:

$$\frac{U(z)}{U_0} = \left[4z(1-z)\right]^{1/m}. \tag{3.15}$$

Here $U_0$ denotes the fluid's longitudinal velocity in the middle of the channel, and parameter $m$ depends on the flow mode. The value $m = 1$ corresponds to laminar flow, the value of $m$ for turbulent flow varies with Reynolds number approximately from 5 to 10. Since large values of $m$ mean profiles with too abrupt fall of the velocity near the wall, hereinafter $m = 5$ will be used as a rule for clearness of pictorial presentations.

Note, this troublesome state of the flow velocity problem means that the turbulence model being in use does not quite correspond to the real processes in the flow. The problem will be discussed in the next chapter. For the present we will continue consideration of the subject matter.

Now, applying sequentially operations "grad" and "curl" to expression (3.15) results in the following formulas:

$$\mathrm{grad}\frac{U(z)}{U_0} \to \left\{ 0,\ 0,\ \frac{4^{1/m}}{m}\frac{(1-2z)}{\left[z(1-z)\right]^{\frac{m-1}{m}}} \right\}, \tag{3.16}$$

$$\mathrm{curl}\frac{U(z)}{U_0} \to \left\{ 0,\ \frac{4^{1/m}}{m}\frac{(1-2z)}{\left[z(1-z)\right]^{\frac{m-1}{m}}},\ 0 \right\}. \tag{3.17}$$

Comparison of these expressions shows that vectors of the velocity gradient and vorticity are of the same value as well as distribution within the flow cross-section. The only difference between them concerns their orientations accordingly along the z- and y-axis directions.

The distributions of the velocity $U(z)/U_0$, velocity gradient $\nabla U(z)/U_0$ and vorticity $\Omega(z)/U_0$ are demonstrated in Fig. 3.1 for the laminar ($m = 1$) and turbulent ($m = 5$ and 100) flows. Some diagrams of interaction between the fluid translational motion and vorticity for the flow parameters

$m = 1$ and $m = 5$ are shown in the figure as well. The interaction diagrams have been already depicted by J. Lighthill [38], whence they are reproduced with some modifications.

As seen from Fig. 3.1, the vorticity $\Omega/U_0$ along with the velocity gradient $\nabla U/U_0$ at the largest values of the parameter $m$ are localized within some narrow area near the wall, known as the boundary layer. At the same time, the gradient and vorticity in the main part of the flow practically vanish, and therein the flow can be considered as the potential. The diagrams in the lower row of Fig. 3.1 illustrate interaction between the two modes of the fluid motion.

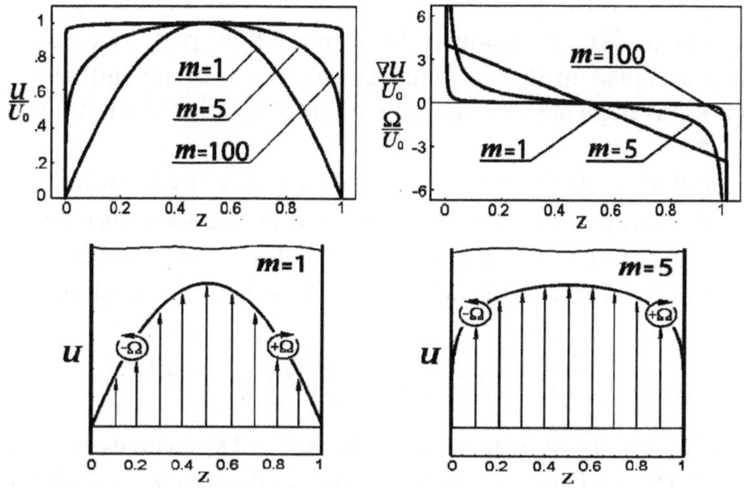


**Fig. 3.1.** Profiles of the velocity, gradient velocity, vorticity, and diagram of their interaction

## 3.2 Hypothesis on pressure profile in a flow cross-section

According to Thomson's theorem, the velocity circulation along some closed contour that encircles a stream line at any point remains constant all along the stream line, and the vorticity moves with the fluid. If the vorticity curl$\boldsymbol{U}$ is zero at some point of a stream line, it must be zero at any other point of this line. Therefore, the vorticity cannot penetrate into the potential part of the flow from that area in which the velocity gradient and vorticity are localized ([3], §9). But for all that a fundamental problem arises as to how the parameter perturbations (or fluctuations) appear in the potential part of the flow, and what is the mechanism of the phenomenon.

We investigate this problem by the example of simple motion of incompressible viscous fluid that obeys the set of equations (2.17). It is more convenient to represent the equations in components for that purpose, implying a steady two-dimensional flow as follows:

$$\left.\begin{aligned}
&\frac{\partial U}{\partial x}+\frac{\partial W}{\partial z}=0,\\
&U\frac{\partial U}{\partial x}+W\frac{\partial U}{\partial z}=-\frac{1}{\varrho}\frac{\partial P}{\partial x}+\nu\left(\frac{\partial^2 U}{\partial x^2}+\frac{\partial^2 U}{\partial z^2}\right),\\
&U\frac{\partial W}{\partial x}+W\frac{\partial W}{\partial z}=-\frac{1}{\varrho}\frac{\partial P}{\partial z}+\nu\left(\frac{\partial^2 W}{\partial x^2}+\frac{\partial^2 W}{\partial z^2}\right).
\end{aligned}\right\} \tag{3.18}$$

It is first worth considering a generally accepted approach to the flow description [2, 3]. As is evident from the graphs shown in Fig. 3.1, small values of the velocity gradient $\partial U/\partial z$, as well as small tangential stress proportional to $\partial U/\partial z$, reside in the central flow part. According to the technique accepted in these circumstances, the viscosity influence on the fluid motion is neglected for the sake of problem simplification in this area of the flow.

In view of the fluid motion's preferred $x$-axis direction, i.e., $W \ll U$, the partial derivative $\partial P/\partial x$ can be replaced by the total derivative $dP/dx$. Taking into account this assumption enables us to rewrite the second equation of the system (3.18) in the form of Bernoulli's equation:

$$\frac{dP}{dx}\approx-\varrho U\frac{dU}{dz},\ \text{or}\ P+\varrho\frac{U^2}{2}\approx const. \tag{3.19}$$

Relative weight of every term in the set (3.18) changes with approach to the wall. Derivative $\partial U/\partial z$ grows, whereas the term $\partial^2 U/\partial x^2$ tends to be small as compared with the second derivative on $z$, and it can be omitted. Now, estimating the ratio of the pressure gradient in the $z$-axis direction to the gradient in the $x$ direction, and comparing the second and third equations of the set (3.18), we shall have the following proportion

$$\left(\partial P/\partial z\right)/\left(\partial P/\partial x\right)\sim W/U\,. \tag{3.20}$$

The velocity component $W$ has just been estimated as small, $W \ll U$. Therefore, the gradient $\partial P/\partial z$ can be considered as small as well with the same accuracy. Then, just as in the case of the main part of the flow, the gradient $\partial P/\partial z$ may be neglected near the wall too, and we obtain for the first approximation:

$$\frac{\partial P}{\partial z}=0. \tag{3.21}$$

These assumptions and replacements rearrange the equation set (3.18) into Prandtl's equations for the laminar boundary layer. The set of the equations follows:

$$\left.\begin{aligned} &\frac{\partial U}{\partial x}+\frac{\partial W}{\partial z}=0, \\ &U\frac{\partial U}{\partial x}+W\frac{\partial U}{\partial z}-\nu\frac{\partial^2 U}{\partial x^2}=-\frac{1}{\varrho}\frac{dP}{dx}. \end{aligned}\right\} \tag{3.22}$$

Note, that here the pressure gradient $dP/dx$ is determined by the longitudinal pressure distribution in the main part of the flow (3.19), and the boundary conditions look like:

$$U\big|_{z=0}=W\big|_{z=0}=0, \quad U\big|_{z\to\infty}\to U(x). \tag{3.23}$$

Applying the set of equations (3.22) for the boundary layer to the problem of flow past a flat half-infinite plate, $x \geq 0$, allows us to obtain the problem's exact solution. It is assumed in the problem (Blasius' problem), that the incident flow velocity $U_0$ is constant, and the velocity component $U$ above the plate increases from zero at its surface to the definite value $U_0$, i.e., $U|_{Z\to\infty}\to U_0$.

Since the velocity $U_0$ in the incident flow is constant by the problem terms, and the pressure in the flow cross-section is accepted as being constant (3.21), the approximate equality $dP/dx \approx 0$ follows from equation (3.19) with the same accuracy, as in the previous assumption. Under these conditions (emphasize the approximate equality $dP/dx \approx 0$) the velocity components $U(x,z)$ and $W(x,z)$ can be expressed from the equation of continuity in (3.22) through the flow stream function $\psi(x,z)$:

$$\left.\begin{aligned} U(x,z)&=\partial\psi/\partial z, \\ W(x,z)&=\partial\psi/\partial x. \end{aligned}\right\} \tag{3.24}$$

Let us introduce the dimensionless stream function $f(\zeta)$, so that

$$\psi=\sqrt{\nu x U_0}\, f(\zeta), \tag{3.25}$$

where $\zeta$ is dimensionless distance to the wall,

$$\zeta=z\sqrt{U_0/\nu x}. \tag{3.26}$$

Then the velocity components (3.24) and the friction force (3.9) acting on a unit of the plate surface are expressed through the dimensionless variables:

$$\left.\begin{aligned} U&=\frac{\partial\psi}{\partial z}=\frac{\partial\psi}{\partial\zeta}\frac{\partial\zeta}{\partial z}=U_0 f'(\zeta), \\ W&=-\frac{\partial\psi}{\partial x}=\frac{1}{2}\sqrt{\frac{\nu U_0}{x}}\left(\zeta f'(\zeta)-f(\zeta)\right). \end{aligned}\right\} \tag{3.27}$$

$$\sigma_{xz} = \eta \left.\frac{\partial U}{\partial z}\right|_{z=0} = \varrho \sqrt{\frac{\nu U_0^3}{x}}\, f''(0). \tag{3.28}$$

These relations show that the velocity components and the friction force depend on the stream function $f(\zeta)$ and its derivatives (see [3], §39). The velocity component $U$ is determined by the first derivative $f'(\zeta)$, the component $W$ depends on difference $\zeta f'(\zeta)-f(\zeta)$, and the friction force is a function of the second derivative $f''(\zeta)$. Substituting relations (3.27) into the second equation of the set (3.22) results in the ordinary differential equation for the stream function $f(\zeta)$,

$$2f''' + f\,f'' = 0\,, \tag{3.29}$$

with the boundary conditions

$$f(\zeta = 0) = 0, \quad f'(\zeta = 0) = 0, \quad f'(\zeta \to \infty) \to 1\,. \tag{3.30}$$

Solution of the nonlinear differential equation (3.29) of the third degree with boundary conditions (3.30) implies some small calculating complexity caused by the requirement at the boundary $\zeta \to \infty$. In essence, its solution needs determination of such a value of the derivative $f''(\zeta)|_{\zeta=0}$, which would satisfy the requirement $f'(\zeta)|_{\zeta\to\infty} \to 1$.

Various skilful analytical or numerical methods were earlier used for this purpose (see [2]). The modern computer-based mathematical systems allow to determine easily the sought for derivative quantity $f''(\zeta)|_{\zeta=0}$. In the given research the interactive system Mathematica (Wolfram Research, Inc) has been used to search out the value. The functions under consideration are computed in *Mathematica* by the following simple program. The outcome of the computation is shown in Fig. 3.2.

**Program 3.1** Evaluation of function $f(\zeta)$ and its derivatives

```
solv_0=NDSolve[{2f'''[ζ]+f[ζ]f''[ζ]=0,f[0]=0,f'[0]=0,
        f'' [0]=0.33},f[ζ],{ζ, 0, 7)}];
solv_1=NDSolve[{2f'''[ζ]+f[ζ]f''[ζ]=0,f[0]=0,f'[0]=0,
        f''[0]=0.33},f'[ζ],{ζ, 0, 7)}];
solv_2=NDSolve[{2f'''[ζ]+f[ζ]f''[ζ]=0,f[0]=0,f'[0]=0,
        f''[0]=0.33},f''[ζ],{ζ, 0, 7)}];

gr_0=Plot[(1/4) Evaluate[f[ζ]/solv_0],{ζ, 0, 7},
        Frame→True, PlotRange→All];
gr_1=Plot[Evaluate[f'[]/solv_1],{ζ, 0, 7}, Frame→True];
gr_2=Plot[Evaluate[f''[ζ]/solv_2],{ζ, 0, 7}, Frame→True];
gr_S=Plot[Evaluate[(1/2) ζ f'[ζ]/solv_1]–
        –Evaluate[(1/2)f[ζ] /solv_0],{ζ, 0, 7}, Frame→True];
```

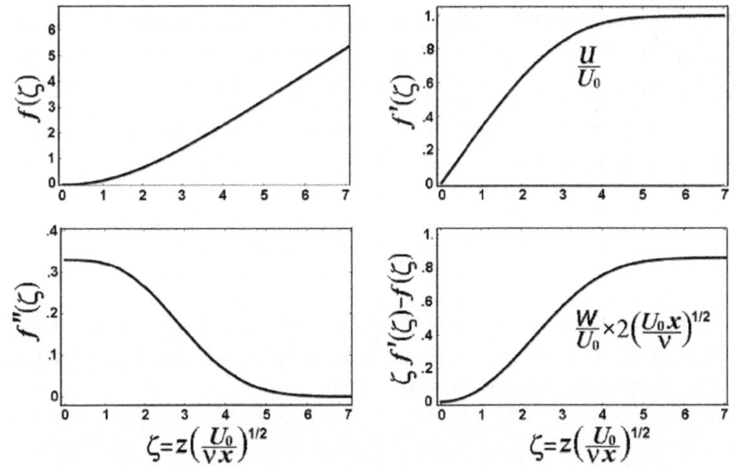


**Fig. 3.2.** Stream function $f(\zeta)$, its derivatives and their combination

The operations solv_n in the program solve equation (3.29) for functions $f^{(n)}(\zeta)$, and the procedures gr_n draw the corresponding graphs. Results of the computations presented in Fig. 3.2 give an uncommon opportunity to estimate the correctness of those assumptions that have been accepted for derivation $f'(\zeta)$ of the equation set (3.29), and of equation (3.22) in the long run.

The graph of the function $f'(\zeta)$ in Fig. 3.2 shows that the velocity longitudinal component $U(\zeta)/U_0$ being equal to zero on the wall surface, $\zeta = 0$, fast increases, so that at $\zeta = 3$ it attains 0.85. After that the augmentation progressively decreases, and the ratio $U/U_0$ tends to 1. When the dimensionless distance from the wall amounts to $\zeta = 5$, the ratio differs from unity by about ~0.2%. If the distance is identified with the edge of the boundary layer, where the friction is practically unimportant, we obtain from the first equation in (3.27) the conditional thickness of the layer, $z = \delta$:

$$\delta = 5\sqrt{\frac{\nu x}{U_0}} = 5\,x\,\mathrm{Re}_x^{-1/2}. \tag{3.31}$$

As an exactly definable characteristic of the boundary layer thickness, the so-called displacement thickness δ* is often used. The thickness δ* corresponds to the distance by which the fluid flow is pushed aside from the plate surface because of the retardation of the stream near the wall.

$$\delta^* = \int_0^\infty \left(1 - \frac{U}{U_0}\right) d\zeta. \tag{3.32}$$

Substituting hither the relation for $U(\zeta)$ from equation (3.27) gives the displacement thickness $\delta^* = 0.34\delta$.

An interesting point in the concept is worth paying attention to. The stream displacement implies fluid transverse motion. This motion is confirmed by the function $W(\zeta)$ behavior, shown in Fig. 3.2. The transverse velocity $W(\zeta)$ in the boundary layer tends, as $z \to \infty$, not to zero but to some non-zero value ([3], §39):

$$\frac{W}{U_0} \approx 0{,}43 \sqrt{\frac{\nu}{U_0\, x}} = 0{,}43\ \mathrm{Re}_x^{-1/2}. \tag{3.33}$$

The fluid transverse motion does not cease even at the outer edge of the friction region, and at the conditional edge of the boundary layer the displacement is a significant quantity still.

## 3.3 Correction of the pressure profile

The principal cause underlying the displacement is obviously the viscous friction proportional to the second derivative $f''(\zeta)$. The derivative is seen from Fig. 3.2 to decrease from its maximum on the wall surface to nearly zero at the conditional edge $\zeta = 5$ of the boundary layer. Notice, the fluid may move in the direction of the $z$-axis, only in the presence of some force acting in this direction. Such a force in a similar fluid, liquid or gas, is the pressure gradient. Therefore, the flow deceleration due to viscous friction should lead to some increase in pressure near the wall.

Hence the first approximation written as the requirement (3.21) seems to be not quite adequate. In this connection, it makes sense to return to formulas (3.27) and (3.22) for the velocity components $U(\zeta)$ and $W(\zeta)$. The formulas have appeared as a solution of equations (3.18) at the supposition that $\partial P(z)/\partial z = 0$ which in the dimensionless variables is written as $\partial P(\zeta)/\partial\zeta = 0$.

It is obvious that the requirement $\partial P/\partial z = 0$ has changed all the set of equations (3.18). First of all, the equation of fluid motion in the $z$-axis direction, third equation in (3.18), has been replaced by simple equality $\partial P/\partial z = 0$. Secondly, one of the items depending on the viscosity has been removed from the equation of motion in the $x$-axis direction. Besides, the pressure distribution $P(x,z)$ has been replaced by the external function $P(x)$, as the first approximation, which is the solution of Bernoulli's equation (3.19) generally pertaining to the main part of the flow.

As a result, only two unknown functions, $U$ and $W$, have remained instead of three functions, $U$, $W$ and $P$. Therefore, the second approximation of the function $P(x,z)$ may hardly be evaluated by the method of successive approximations. Following this method, expressions for $U(\zeta)$

and $W(\zeta)$ from equation (3.27) should be substituted in equations (3.18) to evaluate $P(x,z)$.

To be convinced of this remark's validity, we will substitute the velocity components $U(\zeta)$ and $W(\zeta)$ in the third equation of the original set (3.18). Using the dimensionless variables (3.24–3.28) and Reynolds number with the length $x$ as a scale, on the form $\mathrm{Re}_x = xU_0/\nu$, we obtain the following expression:

$$\frac{1}{\varrho U_0/2}\frac{\partial P}{\partial \zeta}=\frac{1}{\mathrm{Re}_x}\left[\frac{f}{\mathrm{Re}_x}+\left(1-\frac{\zeta^2}{\mathrm{Re}_x}\right)(f''+\zeta f''')\right], \qquad (3.34)$$

In a similar way, the velocity components $U(\zeta)$, $W(\zeta)$ from equation (3.27) are used in Poisson's equation (2.21), that gives the dimensionless formula

$$\frac{1}{\varrho U_0/2}\frac{\partial P}{\partial \zeta}=\frac{1}{\mathrm{Re}^2}\left[\left(\frac{\zeta^2}{4}+\mathrm{Re}\right)(f''+\zeta f''')\right]. \qquad (3.35)$$

The pressure gradients (3.34) and (3.35) in relative values as functions of $\zeta$ are accordingly depicted in Figs. 3.3 and 3.4 for various values of Reynolds number $\mathrm{Re}_x$.

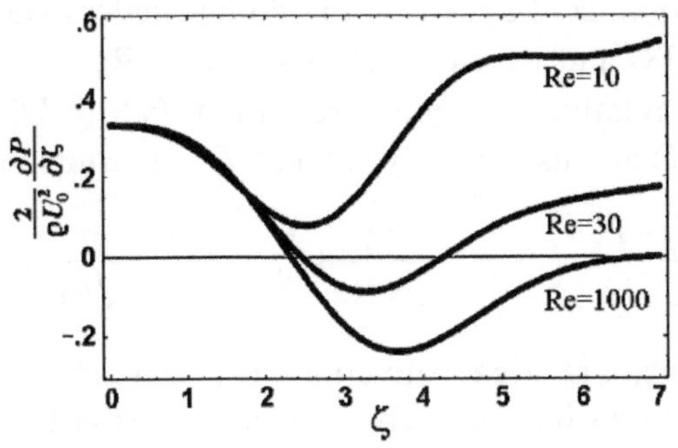


**Fig. 3.3.** Distributions of pressure gradient (3.34) for $\mathrm{Re}_x$ = 10, 30 and 1,000

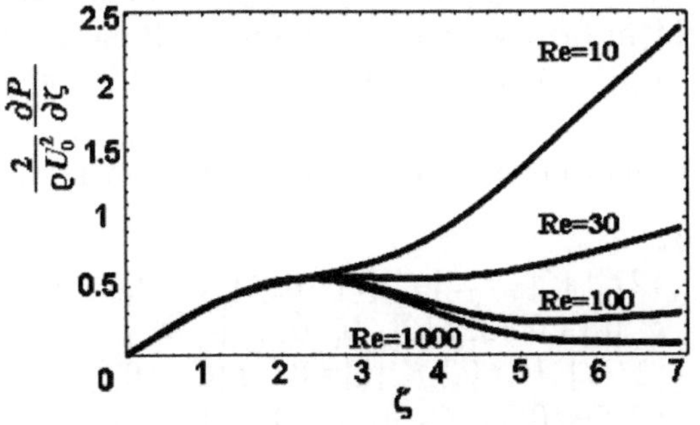


**Fig. 3.4.** Distributions of pressure gradient (2.35) for $\mathrm{Re}_x$=10, 30, 100 and 1,000

The two figures show that the formulas (3.34) and (3.35) predict an unusual pressure profile near the wall and its incomprehensible dependence on its Reynolds number. Evidently, the pressure distribution properties demonstrated by the figures do not agree at all with the velocity function $W(\zeta)$. Also, any physical interpretation of those features can not be found.

Thus we see that the attempts to obtain a suitable kind of distribution $P(z)$, based on equations (3.22), have failed. As has been said, the reason lies in principal distinction of the differential equations (3.22) and (3.18). In the circumstances it is worth using Bernoulli's equation (3.19) in its integral form. Generally, an important preference of an equation in integral form over the differential form consists in a result less sensitive to the faults of such functions as $U(\zeta)$ and $W(\zeta)$.

In equation (3.19) the value of *const* does not change all over the flow, only in the adiabatic case. Otherwise it remains invariable solely along a stream line ([3], §9). The values inherent to various stream lines differ one from another due to dissipation induced by viscous friction. Taking into account the difference and the functions $U(z)$ and $W(z)$, equation (3.19) will be written in the following modified form:

$$\frac{1}{\varrho}P_0 + \frac{U_0^2}{2} \approx \frac{1}{\varrho}P + \frac{U^2}{2} + \frac{W^2}{2} + \nu\frac{\partial U}{\partial z}. \qquad (3.36)$$

Now the equation can be rewritten in dimensionless variables. For that formulas (3.24–3.28) and Reynolds number Re = $\mathrm{Re}_x$ = $xU_0/\nu$ are to be used. Expressing relative overpressure $(P(\zeta)-P_0)/(\varrho U_0/2)$ through the stream function $f(\zeta)$ and its derivatives $f'(\zeta), f''(\zeta)$ results in the equation

$$\frac{P(\zeta)-P_0}{\varrho\, U_0/2} = 1 - f'^2 - \frac{1}{4\,\mathrm{Re}}(\zeta f' - f)^2 - \frac{2}{\sqrt{\mathrm{Re}}} f''. \qquad (3.37)$$

The non-dimensional stream function $f(\zeta)$, and also its first $f'(\zeta)$ and second $f''(\zeta)$ derivatives have already been computed in Program 3.1. Now the relative overpressure $(P(\zeta)-P_0)/(\varrho U_0/2)$ can be determined by choosing the flow cross-section $x$, and by indicating Reynolds number $\mathrm{Re}_x$ = $xU_0/\nu$, where $U_0$ is the incident flow velocity over a plate. Program 3.2 is intended for the overpressure estimation in the flow for $\mathrm{Re}_x$ = 30, 100 and 1,000.

**Program 3.2** Overpressure in the three cross-sections of a flow over plate

```
Re={30, 100, 1000};
Solve_0=NDSolve[{2f'''[ζ]+f[ζ]f''[ζ]=0,f[ζ] = 0,
                f''[ζ]=0.33,f[ζ], {ζ, 0, 7}];
Solve_1=NDSolve[{2f'''[ζ]+f[ζ]f''[ζ]=0, f[ζ] = 0,
                f''[ζ]=0.33,f'[ζ] , {ζ, 0, 7}];
Solve_2=NDSolve[{2f'''[ζ]+f[ζ]f''[ζ]=0, f[ζ] = 0,
```

```
            f''[ζ]=0.33, f''[ζ] , {ζ, 0, 7}];
gi[[i]] = Plot[1–Evaluate[f'[ζ]²/.solve_1]–1/(4Re[[i]])×
   ×{Evaluate[ζf'[ζ]²/.solve_1]–Evaluate[f[ζ]/.solve_0]}²–
   –Evaluate[(2/Re)f''[ζ]/.solve_2], {ζ,0,7}, Frame→True,
   PlotRange→{{0,7},{0,1}},FrameLabel→{"ζ", "2(P–P₀)/ϱU²"},
   DisplayFunction→Identity];
gtxt=Graphics[{Text["Re=30",{3.8,0.88},{0,0}],
  Text["Re=100",{3.9,0.68},{0,0}],Text["Re=100'0',{4,0.48},{0,0}]],
comb=Show[{gi[[1]], gi[[2]], gi[[3]], gtxt},
        DisplayFunction→&DisplayFunction;
```

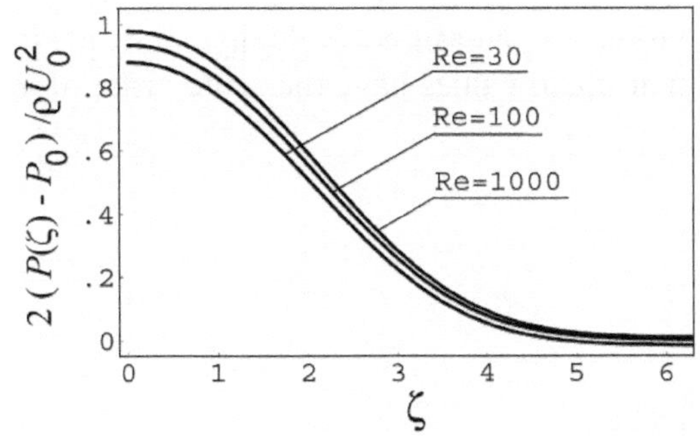


**Fig. 3.5.** Overpressure in flow cross-section for $\mathrm{Re}_x$=30, 100 and 1,000

In the program, just the same as in Program 3.1, operations solv_n determine functions $f^{(n)}(\zeta)$ by solving equation (3.37). After that, the procedures gi[[i]] and gtxt prepare data for construction of the required graphics with suitable inscriptions. The command comb draws images of the relative overpressure as functions of the parameter $\zeta$. The result of the computations for the indicated values of Reynolds number are shown by the graphics in Fig. 3.5.

The findings represent the overpressure in the form of monotone functions that have no singularities. Also the functions demonstrate a reasonable transition from the boundary layer to the main potential part of the flow, and show acceptable dependence on Reynolds number $\mathrm{Re}_x$. Thus, in the beginning we have started with the assumption that the pressure distribution across the flow does not depend on fluid current viscosity. According to this understanding, the requirement (3.21), $\partial P/\partial z = 0$, has been accepted, though as a first approximation. However, further consideration has shown that the requirement (3.21) needs some correction. The corrections make a distinction in kind in the phenomenon interpretation.

The viscosity taken into account has led to equation (3.37) that results in pressure dependence on the coordinate $z$, represented by the diagram in Fig. 3.5. As follows from the diagram, the pressure in the region of the wall friction and flow deceleration increases. The maximum is located on the wall surface, which shows the best correlation with the velocity profiles (3.15) and the velocity gradient distributions (Fig. 3.1) observed in experiments.

The correction introduced into the relative overpressure distribution has an order of magnitude $\varrho U^2/2$ which is usually small as compared with the pressure $P_0$ in the flow. When $M \ll 1$, it comes to several percents from the pressure $P_0$. Indeed, in water ($\nu = 0.01$ cm$^2$/s) moving along a flat wall with velocity, say, of 20 cm/s, the pressure correction is estimated as $\varrho U^2/2P_0 \approx 0.02$ with the other parameters being $P_0 = 1$ atm, $x = 1$ cm, $\mathrm{Re}_x = 10^4$.

It should be emphasized in this connection that the field of our consideration relates namely to moderate flows, $M \ll 1$. Under such conditions the ratios of measured velocity and pressure fluctuations (oscillations) to their mean values have the same order of magnitude.

# 4
# Perturbations in viscous flow

## 4.1 Fluid motion from the start

Kinetic energy of an incompressible fluid motion in volume V is

$$E=(\varrho/2)\int U^2 d\mathrm{V}.$$

Fluid viscosity involves energy dissipation due to friction, which is defined by the time derivative $\partial E/\partial t$. By means of Navier–Stokes' equation and some mathematical manipulation, this derivative can be written in a form of the following expression [3, §16]:

$$\frac{\partial E}{\partial t}=-\frac{\eta}{2}\int\left(\frac{\partial U_i}{\partial x_k}+\frac{\partial U_k}{\partial x_i}\right)^2 d\mathrm{V}. \tag{4.1}$$

When a viscous incompressible fluid moves along a wall by length $L$, and width Y, the layer by height $z$ looses the energy

$$\frac{\partial E}{\partial t}\approx-2\eta\mathrm{Y}\int_0^z\int_0^L\left(\frac{\partial U}{\partial z}\right)^2 dx\,dz. \tag{4.2}$$

The derivative $\partial U/\partial z$ grows rapidly at the wall surface, and according to formula (4.2), the main part of the energy losses are located near the walls of the channel.

Some outside force and energy source should exist to compensate for the losses and maintain the fluid motion in a constant state. The fluid (liquid or gas) is set in motion by pressure gradients within a simple wave (Riemann's wave) [3, §101]. There are three ways to excite a simple wave in channels [38]: by pressure drop between a receiver and the channel's open end, by outflow due to a great volume, and, lastly, by a moving piston.

Disturbances at the channel entrance in the first two alternative cases complicate unpredictably the analysis of the flow properties [25]. Therefore, it is better to decide in favour of the variant of the fluid motion in front of



L.N. Pyatnitsky, *Turbulence Nature and the Inverse Problem*, Fluid Mechanics and Its Applications 89, 43–53.

the moving piston, where flow parameters can be calculated, provided that the piston advances at rather moderate velocity.

In the beginning we will ascertain features of one-dimensional motion. Without loss of generality, the motion is more convenient for exploring in gases, as the equation of state for gases has a simple form. Solution of the problem concerning the one-dimensional motion of gas in front of the piston (without any friction) is well known (see, e.g. [3], §§99, 101).

Let us assume that a piston starts moving at the moment $\tau = 0$ from the point $x = 0$ in the direction $x > 0$ according to some law $X(\tau)$. While moving, it generates a sequence of elementary pressure waves. On the whole the waves present a simple wave traveling in front of the piston. Dependence of the fluid velocity at a current position $x$ and running time $t$, $U(x,t)$ within the simple wave is determined by parametric equations

$$\left.\begin{aligned} U &= X'(\tau), \\ x &= X(\tau) + (t-\tau)\,[c_0 + \frac{\gamma+1}{2} X'(\tau)]. \end{aligned}\right\} \tag{4.3}$$

Here $c_0$ denotes the sound velocity in the gas under consideration at quiescent state, and $\gamma$ is the gas adiabatic exponent. As these equations show, the velocity $U$ in the simple wave is a variable value. Along with the velocity $U$ vary the pressure $P$, local sound velocity $c$, and also the gas density $\varrho$:

$$\left.\begin{aligned} \frac{P}{P_0} &= \left(1 + \frac{\gamma-1}{2}\frac{U}{c_0}\right)^{\frac{2\gamma}{\gamma-1}}, \\ c &= c_0 + \frac{\gamma-1}{2} U, \\ \frac{\varrho}{\varrho_0} &= \left(1 + \frac{\gamma-1}{2}\frac{U}{c_0}\right)^{\frac{2}{\gamma-1}}. \end{aligned}\right\} \tag{4.4}$$

As the elementary pressure wave propagates at the new velocity $U+c$ which depends now on the variable density, and is different for various values $x$, the simple wave shape is deformed in time. Let the piston velocity $X'(\tau)$ increase at first, and then tend to the limit $U_0$ with the characteristic time $\theta$. To be definite, it is represented by the expression

$$X' = U_0\left(1 - \exp(-\tau/\theta)^{\mu}\right), \tag{4.5}$$

where the exponent $\mu$ can be supposed equal, for instance, $\mu = 2$.

The spectrum of the piston motion and of the waves radiated by the piston depend evidently on the frequency $\omega = 2\pi/\theta$. The spectrum density of the velocity (4.5) variable component looks like

$$F(\omega)=\left(\theta/\sqrt{2}\right)\exp(-\theta^2\omega^2/4), \tag{4.6}$$

and the spectrum of the constant component is expressed through the Dirac delta function $\sqrt{2\pi}\,\delta(\omega)$.

Inevitable natural irregularity of the piston motion, omitted in the formula (4.5), can be included as an additional term. This term can be presented by a harmonic component of the frequency $\omega = 2\pi n/\theta$ and some small amplitude $g \ll 1$:

$$\delta X' = U_0\, g \sin\left(2\pi n \tau/\theta\right). \tag{4.7}$$

Evolution of the overpressure function $\Delta P(x)/P_0 = (P(x)-P_0)/P_0$ in the simple wave, similar to the velocity $U(x)$ evolution, can be demonstrated by comparison of the function computed for different instants and various velocities of the piston. The function $\Delta P(x)/P_0$ at different values $U_0$ is evaluated by equations (4.3) where the parameters are given by expressions (4.4) and (4.5). Inasmuch as the ratio $U(x)/c_0$ is always small, expressions for pressure and density in equation (4.4) can be expanded into series and replaced by the approximated values including only the first terms of the expansions, e.g., $P(x)/P_0 \approx 1 + \gamma U(x)/c_0$.

In the beginning the ratio $\Delta P(x)/P_0$ is computed as a function of the final velocity $U_0$ for the piston moving according to the law (4.5). Then the function for the same piston motion (4.5) is corrected for the irregularity (4.7). The piston motion with the irregularity (4.7) seems to be closer to real system behavior. To compare the two modes of the piston motion, the following parameters are employed in these mental experiments: the piston final rate $U_0 \in [5,20]$ m/s, sound velocity $c_0 = 350$ m/s, characteristic time $\theta = 0.002$, time instants $t = 3\theta, 10\theta$, and $n = 5$, $g = 0.05$.

Further the program for computation of the function $P(x)/P_0$ follows. In the system *Mathematica* the program is fulfilled with the help of an additional package addressed as <<Graphics` ParametricPlot3D`.

**Program 4.1** Pressure distribution in a simple wave

```
c=350; θ=0.002; t1=3θ; t2=10θ; γ=1.4;
X'=U0(1-Exp[-(t/θ)^2]);
X=Integrate[X',{t,0,t}];
ΔP/P0=100γX'/c;
x3θ=X+(t1-t)(c+X'(γ-1)/2);
x10θ=X+(t2-t)(c+X'(γ-1)/2);

g3θ=ParametricPlot3D[{x3θ,X',ΔP/P0}, {t,0,t1},{U0,5,20,2.5},
      PlotRange→{{0,8.5}, {5,20.1},{0,9}}, BoxRatios→
      {1,1,0.7},AxesLabel→{"x.m", "U0, m/s", "ΔP/P0,%"},
```

```
          ViewPoint→{7,–11,1}, DisplayFunction→Identity];
g10θ=ParametricPlot3D[{x10θ,X′,ΔP/P0}, {t,0,t2},{U0,5,20},
          PlotRange→{{0,8.5}, {5,20.1},{0,9}}, BoxRatios→
          {1,1,0.7},AxesLabel→{"x.m", "U0, m/s", "ΔP/P0,%"},
          ViewPoint→{7, –11,1},DisplayFunction→Identity];
comb=Show[{g3θ,g10θ},DisplayFunction→$DisplayFunction];
```

The first line of Program 4.1 includes information on the chosen values of the problem's initial parameters. The next five lines stand for the formulas of the parametric equations that calculate the piston motion and overpressure distributions along the simple wave, *x*3θ and *x*10θ, corresponding to the profiles at two instants, t1 = 3θ and t2 = 10θ. The blocks g3θ and g10θ prepare data for the graphic representation.

In the end, the operation named as 'comb' constructs the graphs of the overpressure distributions along the simple wave length *x* for two given instants, that are shown by the graph *a* in Fig. 4.1. A similar solution for the same initial data, but with correction (4.7), where $n = 5$ and $g = 0.05$, is shown as a graph *b*.

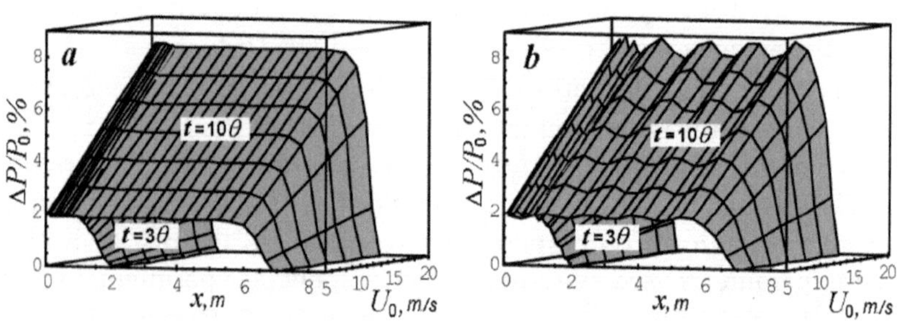


**Fig. 4.1.** Evolution of pressure in a one-dimensional simple wave

As evident from the graphs in Fig. 4.1, the pressure in the simple wave (as well as the flow velocity $U$ similar to it) settles at the front for a short time, and then remains constant along the remainder of the simple wave. The front formation in space and time depend obviously on peculiarity of the piston acceleration development, i.e., on the kind of function (4.5). The irregularity in the velocity excites the pressure oscillations that constitute perturbations moving within the simple wave as a train of acoustic wave packets.

The pressure in the simple wave is proportional to the piston's final velocity $U_0$, provided that the piston motion follows the law (4.5), with $U_0$ being constant. When the piston motion law varies, the pressure distribution in the wave follows the variations. In particular, graph *b* in Fig. 4.1 shows that the piston velocity oscillations lead to similar oscillations in the pressure distribution. Another example of practical importance relates to the piston velocity's permanent growth, $U_0(t)$. In this case the pressure in the simple wave becomes similar to its distribution in the extended wave front, gradually increasing from $P_0$ to the maximum quantity which responds to the highest value of the velocity $U_0$.

As mentioned above, a fixed point in the simple wave travels at the velocity ([3], §101)

$$\left(\frac{\partial x}{\partial t}\right)_U = U + c(U), \tag{4.8}$$

or, after integration, according to the trajectory

$$x = [U + c(U)]t + F(U),$$

where $F(U)$ is a function similar to equation (4.5), and $c(U)$ is given by the relation $dU/d\varrho = c/\varrho$. Hence, the total velocity of the wave packet may be regarded as the result of superposition of the packet propagation relative to the fluid at the sound velocity $c$ on the movement of the fluid itself at the variable velocity $U$.

The dependence (4.8) leads ultimately to a discontinuity, or a shock wave, formation at the simple wave front, and at the front of every wave packet as well. Analytically the place and time of the discontinuity formation is determined by two simultaneous equations:

$$(\partial x/\partial U)_t = 0, \ (\partial^2 x/\partial U^2)_t = 0. \tag{4.9}$$

When the simple wave front moving in the fluid at rest is considered, the first requirement vanishes, and the second one holds at the condition $U = 0$:

$$(\partial x/\partial U)_t\Big|_{U=0} = 0$$

In the case of an arbitrary pressure amplitude, the simple wave turns into a shock wave as far as the discontinuity is formed, being accompanied by a fluid entropy jump, while the entropy in the simple wave is constant. However, if $U_0 \ll c$ and the pressure amplitude is small, the change in entropy at the discontinuity is of the third order of smallness. So that the wave with the discontinuity formed can still be regarded as a simple wave in the second approximation ([3], §102).

## 4.2 Simple wave and wave beam

Generally, the one-dimensional simple wave, and the wave packet as well, is substantially an abstraction, for the effects of a finite aperture in the channel where the wave travels are not considered in equation (4.3). Meanwhile, the channel aperture has a strong impact on flow behavior and structure.

In fact, the fluid flow in a channel means that we deal not with a simple wave or wave packet, but with a wave beam. In addition, the fluid moving within the simple wave contacts with the channel walls. The viscous friction (3.9) originated from the contact creates a pressure gradient $dP/dx$ directed against the current. That leads to gradual deceleration of the flow in the channel down to full immobility.

If an acoustic plane wave damps during some time $T$, then the differential pressure vanishes, $\Delta P/P_0 = 0$, at the distance $c_0T$ covered by the wave. Therefore, the greatest length of the simple wave being driven by the piston is limited by that distance, and the system of the piston-simple wave becomes practically stationary, if, of course, the piston motion is steady, and the channel length is large enough.

The one-dimensional image of a simple wave implies that elementary plane waves transfer the momentum flux in the fluid only in the $x$-axis direction, irrespective of diffraction phenomena. It is the principal distinctive feature of that one-dimensional simple wave. However, in reality the fluid within the simple wave moves in a channel of finite cross-section, say, of diameter $d$.

Under the circumstances the flow is not one-dimensional motion, and it should be considered as a wave beam. According to Rayleigh's law on diffraction divergence, any wave beam diverges, as is illustrated by the schematic diagram in Fig. 10.1. In Chapter 10 we discuss the beam of electromagnetic radiation. In this case we deal with acoustic radiation. But it does not matter, for the principal point consists in the requirement that the radiation has to be of a wave nature [41].

Any parameter of the flow may be brought to conformity with Fourier image, and the fluid motion recognized as a set of waves. Some wave is characterized by its frequency $\omega$, wave length $\lambda$ and wave vector $\boldsymbol{k}$. If the wave vector of the beam in its initial state is $\boldsymbol{k}_0 = 2\pi/\lambda \boldsymbol{s}_0$, where $\boldsymbol{s}_0$ denotes the unit vector directed along the beam axis, and the beam diameter amounts to $d$, then the divergence angle is $\theta \approx \lambda/d$.

Correspondingly, the wave vector in the direction $\theta$ is $\boldsymbol{k}(\theta) = 2\pi/\lambda \boldsymbol{s}(\theta)$, where $\boldsymbol{s}(\theta)$ means the unit vector in the direction of the angle $\theta$. The wave front of the diverging wave looks like the sector of sphere limited by the angle $\theta$. Hence this Rayleigh diffraction divergence appears as additional momentum flux in the radial direction, and a radial component of the elementary waves ([3], §61).

The diffraction phenomena obeying the wave equation are known not to depend on the wave field nature, but on the beam diameter $d$ and wavelength $\lambda$. Nevertheless, the estimation technique of the beam divergence state varies slightly in different areas of knowledge.

For instance, the properties of the optical beams are estimated by the divergence angle $\theta \sim \lambda/d$ ($\lambda = 2\pi c/\omega$). The wave front in Gaussian beams is usually thought of as plane, while the beam length $L$ does not exceed $L \approx \pi d^2/\lambda$ (Rayleigh length). Properties of the beams created by a piston sound radiator are estimated in acoustic applications by the value of the so-called wave parameter $P = (\lambda L/S)^{1/2}$, where $S$ is the surface area of the piston (see, e.g. [42]).

Close to the piston, when, in the so-called near-field region, $P \ll 1$, the wave front is considered as plane. When $P \sim 1$, the wave front is recognized as convex; it tends to a spherical segment. And wide apart, $P \gg 1$, the wave takes a spherical-like configuration. The flow within the simple wave in the example considered above can be characterized by the wavelength $\lambda \sim c\theta \sim 70$ cm. Let the pipe diameter be $d = 3$ cm, or $S \sim 10$ cm$^2$. Then the wave parameter reaches the value of $P = 1$ already at the distance $L < 0.2$ cm, i.e., the wave front obtains the spherical form practically at once.

The plane wave reflection from the walls at oblique incidence is known ([3], §79) to cause intense dissipation. The energy is absorbed near the wall due to the fluid temperature rise at the reflection spot. A fraction of the wave mean energy flux density incident upon a unit area of the wall dissipated near the surface is defined by relation

$$\gamma = \frac{2\sqrt{2\omega}}{c\cos\vartheta}\left[\sqrt{\nu}\sin^2\vartheta + \sqrt{a}\left(\frac{c_p}{c_v} - 1\right)\right], \tag{4.10}$$

where $\vartheta$ is a local incident angle of the wave, $c_p$ and $c_V$ are thermal capacities, $a$ is a factor of the fluid thermal diffusivity.

This important question is worth dwelling on, for any spherical wave propagation in a channel is necessarily accompanied by the wave reflections from the channel walls at various angles. This expression, derived for the reflection of a plane wave, can evidently be accommodated to the reflection of the spherical wave, at least when the wave radius $R$ is not too large.

However, we are interested in the energy absorption $G_x$ at the spherical wave reflection spot while it runs along the wall surface in the direction of the $x$-axis. The wave incidence local angle $\vartheta$ changes along with the wave radius $R$ in this case. Hence, the components $u$ and $w$ of the velocity oscillation amplitude u and the motion velocity of the reflection spot $\chi'$ depend on the radius $R$ as well [37].

The radius is convenient to employ as dimensionless value with the scale being the cross-section characteristic dimension $d$. Then the wave incidence angle $\vartheta(R)$ in these variables is

$$\vartheta = \arccos(R/2), \tag{4.11}$$

and the amplitudes of the oscillation velocity components in longitudinal $u$ and transverse $w$ directions are expressed correspondingly as

$$u = \mathrm{u}(R)\sin\vartheta,\ w = \mathrm{u}(R)\cos\vartheta, \tag{4.12}$$

where $\mathrm{u}(R)$ is the velocity oscillation amplitude within the spherical wave packet,

$$\mathrm{u}(R) \sim 1/R. \tag{4.13}$$

Now the amplitudes of the velocity components are

$$\left.\begin{aligned} u &\sim \sqrt{4R^2-1}/2R^2, \\ w &\sim 1/2R^2. \end{aligned}\right\} \tag{4.14}$$

The velocity of the reflection spot motion is determined by the first derivative from its dimensionless coordinate $\chi = x/d = 1/2\mathrm{tg}\vartheta$ with respect to the wave radius $R$, $\chi' = (d\chi/dR)(dR/dt)$. The differentiation results in the following simple relation

$$\chi' = cR(R^2-1/4)^{-1/2} \tag{4.15}$$

At last, the energy dissipation per unit area of the wall surface can be represented in the form

$$\mathrm{G}_x = \gamma R^{-2}. \tag{4.16}$$

Functions of the dissipation $\mathrm{G}_x(R)$, of velocity amplitude components $u(R)$ and $w(R)$ at the running reflection spot, and the velocity $\chi'(R)$ of this spot are shown in Fig. 4.2.

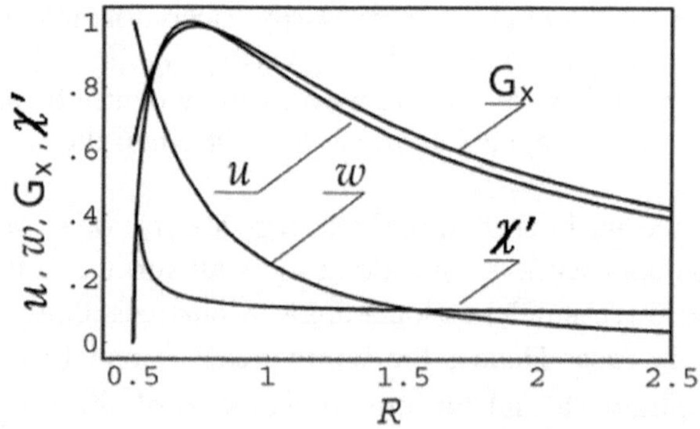


**Fig. 4.2.** Functions $\mathrm{G}_x(R)$, $u(R)$, $w(R)$, $\chi'(R)$ at the moving point of the spherical wave reflection

The constants $\nu \approx a \approx 0.15\ \mathrm{cm^2 s^{-1}}$ and $c_P/c_V = 1.4$ (for air) have been used in evaluation of the functions shown in Fig. 4.2. To combine all the functions on the graph in one vertical scale, they are represented at different factors. The function $\mathrm{G}_x(R)$ is depicted as the ratio of the current value to its maximum, the amplitudes $u(R)$ and $w(R)$ are given in the scales 1:1 and 1:2 correspondingly, the velocity $\chi'(R)$ is presented in ratio 1:10.

## 4.3 Origin of pressure perturbations

It is important to notice some peculiarities of the functions depicted in Fig. 4.2. The velocity component $w(R)$ and the travel velocity of the reflection spot $\chi'(R)$ are monotonic functions, though when $R \to 1/2$, the function $w(R)$ tends to 2, whereas $\chi'(R)$ increases without limit. At the same time, the absorption and the longitudinal oscillation amplitudes, the functions $G_x(R)$ and $u(R)$ correspondingly, have acute maxima.

Locations of the maxima are worthy of special attention, for they correspond to nearly the same spherical wave radii, namely $R_{\mathrm{Gmax}} = 0.73$ and $R_{u\max} = 0.71$, and approximately the same incident angles, $\vartheta_{\mathrm{Gmax}} = 46.7°$ and $\vartheta_{u\max} = 45.2°$. So that the maxima are separated on the wave surface by the very small distance which does not exceed $\Delta x = 0.03$, i.e., we may suppose that locations of the maxima coincide. Also, the amplitudes $u(R_{\max})$ and $w(R_{\max})$ become equal in the region of the coincidence, and the spot of the wave reflection travels at sound velocity, within the calculation accuracy.

Now notice, that formula (4.12) for the velocity amplitude of the longitudinal oscillations within the wave packet is in essence the same as formula (1.12) for the small exciting motion $\psi$ which is used for the analysis of the boundary layer instability. However, there is some distinction. The formulas (4.12) refer to a certain physical process that gives an opportunity to construct the mechanism of the flow disturbances. It is grounded on the following principles.

Generally, the velocity amplitude $u$ of the longitudinal oscillations is small in comparison with the flow velocity component $U$, but both of them are zero on the wall surface. Meanwhile near the wall, within the boundary layer, the velocity $U(z)$ decreases slowly as $z^{1/m}$. It is clearly seen from the formula (3.15) and Fig. 3.2. At the same time, the amplitude $u(z)$ is proportional to the difference of the kind $1-\exp(-z\,(\omega/2\nu)^{1/2})$, and decreases only at the very wall.

Therefore, the velocity $u$ of the fluid's longitudinal oscillations becomes of the order and even more than the flow velocity $U$ in immediate proximity to the wall along the short length $\Delta x$ ([3], §79). Then a considerable oscillating local gradient of the tangent velocity and sequent tangential discontinuity appear in the zone of the boundary layer. The tangential discontinuity is an absolutely unstable structure which just turns into a formation of the pressure jump within the zone ([3], §29). It seems to be the source of that sought exciting motion, or the primary perturbing motion.

The tangential velocity discontinuity and subsequent events develop against the background of heightened pressure near the wall. Indeed, the viscous flow friction creates a pressure increase existing near the wall, considered in §3.3. The spherical wave reflection adds the local brief pressure jump of the value to $p \approx 2c\varrho u^2 \cos^2 \vartheta$ ([3], §66). Besides, the pressure grows at the moving point of the wave reflection due to dissipation of the wave energy, which, as we have seen, occurs in the gradient field of the flow velocity.

Because all these phenomena are concentrated in the field of the viscous flow friction, i.e., within the boundary layer, the radial dimension $a$ of the primary perturbation can be estimated as $a \approx \delta$. At the same time, the listed processes leading to formation of the tangential discontinuity and pressure jump develop during the time $\tau$ of the spherical wave interaction with the boundary layer. As the time $\tau$ is rather small, of the order of $2\delta/c$, the local perturbation may be presented by instantaneous distribution $f(R)|_V$ with parameters:

$$\tau \approx 2\delta/c, \Delta x \approx 0.03, a \approx \delta. \tag{4.17}$$

A perturbation in the form like $f(R)|_V$ propagates over a channel as the acoustic wave packet (2.30). When the volume V is small, i.e., the distance $r$ covered by the perturbation border is getting large, $r/a \gg 1$, the wave packet front takes practically the spherical shape. Surely, the spherical shape is distorted after multiple reflections from the walls.

The perturbations follow one after another, and the sequence of the wave packets, including the primary and reflected ones, forms some structure of perturbations in the channel. Each of the spherical wave packets implies oscillations of hydrodynamic parameters, and their totality is understood as a spatio-temporal field of the fluctuations.

Now, some physical comments might be made on the expression (4.2) which this chapter began with. The fluid in the channel comes to the motion under the action of an outside force produced by the piston in our case. It transmits energy to the fluid by emitted pressure waves. A certain quantity of the power transmitted by the piston is consumed in overcoming the viscous friction. Some part of this quantity is laid out directly for fluid heating and is partially transmitted to the walls due to the thermal conductivity.

The remaining part goes toward generation of the acoustic waves which create a field of perturbations in the form of oscillating parameters, thereby scattering the energy over the flow. While propagating, the waves and oscillations gradually damp owing to viscosity and thermal conductivity, and the absorbed energy turns into heat.

A spherical wave packet reflection from a wall in the presence of some boundary layer had not been investigated in detail till now. The process pattern described above is an evident simplification. Indeed, some factors have fallen out of the consideration, such as the vorticity (2.19), overpressure in the layer of viscous friction (2.21), (3.37), oscillation kinetic energy conversion into potential energy within the reflection region, and destruction of the compression wave front in the boundary layer, which is discussed in the following chapter.

Nevertheless, analysis of the simple wave structure ([35–37], etc.) shows that the important peculiar properties of the pressure wave reflection, connected with the nonzero pressure gradient in the $z$-axis direction, $\partial P/\partial z \neq 0$, and propagation of the primary local perturbation in the form of the spherical wave packet, meet reliable experimental verification.

The experimental part of the study will be considered in Chapter 8. For the present we will explore properties of the primary perturbation propagation in channels.

# 5
# Perturbation in channels

## 5.1 Perturbations in semi-infinite space

As said at the end of the previous chapter, according to equation (4.17), the primary acoustic perturbation arises within a small volume V practically instantaneously. Thus it may be represented by a function like $f(R)|_V$. At the distance $r \gg a$ beyond the edges of the boundary layer, initial configuration of the volume V does not noticeably influence the perturbation wave contour at later stages, and the wave front can be considered to be of a spherical shape.

Let such a perturbation originate in the fluid at rest on a wall in the form of a hemisphere with radius $a$ at an instant $t = 0$. The spatio-temporal field, which the perturbation $f(R)|_V$ creates while propagating, is described by the integral (2.30). But the velocity potential $\varphi$ is well-known (see, for example [3], §70) for the outer problem, $r \notin V$, in the given case of spherical symmetry (see, for example [3], §70). It looks like $\varphi = f(r,t)/r$, where function $f(r,t)$ is the solution of the one-dimensional wave equation

$$\frac{\partial^2 f}{\partial t^2} = c^2 \frac{\partial^2 f}{\partial r^2}. \tag{5.1}$$

The solution of this equation has the simplest form when the initial perturbation $f(R)|_V$ is set by a distribution of the fluid density variable part $\rho(R)$. Since the volume V is small, we can consider the part to be a constant value, $\rho(R) = \Delta = \text{const}$. Then the solution takes the form:

$$\left.\begin{aligned} &\text{if } |r-ct| > a, \text{ then } \rho = 0; \\ &\text{if } |r-ct| \le a, \text{ then } \rho = \frac{\Delta}{2}\frac{r-ct}{r}. \end{aligned}\right\} \tag{5.2}$$

The rendition of the solution (5.2) looks like a spherical wave packet of thickness $2a$. It propagates in the fluid with sound velocity $c$, and the amplitude is proportional to $r^{-1}$. Within the packet, $|r-ct| \le a$, the density



L.N. Pyatnitsky, *Turbulence Nature and the Inverse Problem*, Fluid Mechanics and Its Applications 89, 55–71.

ρ changes from Δ/2 to –Δ/2 in relation to the density average value ϱ. However, if the density distribution in the primary perturbation, $R \leq a$, depends on $R, f(R)|_{R \leq a}$, or on time, $f(c\tau)|_{c\tau \leq a}$, then the factor (*r–ct*) in the product Δ (*r–ct*) should be replaced by the function *f*(*r–ct*) corresponding to the initial distribution of the parameters.

We may conclude from relations (2.26), that the pressure and density functions in the wave packet are similar to each other, for both of them are proportional to the time derivative of the potential, $\partial\varphi/\partial t$. At the same time, the velocity ***u*** is defined by the potential gradient.

In the case of spherical symmetry, the gradient can be evaluated by substituting the derivative $\partial\varphi/\partial t$ into the equation of motion from the system (2.24), and next integrating the result for an arbitrary function *f*(*r–ct*). After the integration, distributions of the parameters (2.26) within the wave packet are represented as follows:

$$\left.\begin{aligned} &p \sim \rho \sim \frac{1}{r} f(r-ct); \\ &\mathrm{u} \sim \frac{1}{r} f(r-ct) - \frac{1}{r^2} \int_{r-ct} f(r-ct)\, d(r-ct)\,. \end{aligned}\right\} \tag{5.3}$$

While propagating in the flow, every spherical wave packet brings along with it some oscillations of the parameters within the packet thickness $2a$. Figure 3.1 and formula (3.15) have shown that when the flow is intensified, i.e., the longitudinal velocity $U$ and the index $m$ increase, the region of the velocity gradient $\partial U/\partial z$ gets thinner, and the gradient itself strengthens. This strengthening leads to more intense formation of the parameter perturbations in the boundary layer. In the beginning, while the velocity gradient is relatively small, the perturbations arise sporadically. But then the perturbations appear more frequently, and the generation process becomes regular.

The sequence of perturbations and superposition of parameter oscillations excited by the wave packets form a spatio-temporal field. The field state is characterized by the alternation or intermittency factor $\gamma$ of the turbulence [2] ([3], §35). When $\gamma < 1$, the case of low frequency of generation, the turbulence is of an intermittent character. When $\gamma = 1$, the case of regular high frequency, the perturbation dynamics is in an equilibrium state, and the turbulence is completely developed.

Given all that, the notion concerning the free surface shape of the boundary layer implies its time-averaged position. The surface's instantaneous shape is highly irregular even if $\gamma = 1$, as verified by experimental data. This irregularity results from the discrete nature of origination of the primary perturbations.

The above reasoning, based on the wave nature of the turbulence, implies a certain wave mechanism of the parameter oscillations in the turbulent flow. However, the wave concept assumes one more mechanism of turbulent oscillations, which is concerned with travel, development and interaction of the vortices in the flow. This mechanism is similar to the one considered by L. Richardson [11].

This mechanism proceeds from the fact that the strong velocity gradient at the wall entails a vortex motion, and the vortices evolve and interact in the flow. However, we see here the problem that was mentioned above, i.e., the vorticity does not penetrate into the main potential part of the flow from the boundary layer, as follows from Thomson's theorem. In addition, the vorticity localization within the boundary layer is in a stable state under the given conditions, according to a finding in ([3], §35).

Now, we have an opportunity to solve the problem in the framework of the wave approach to turbulence. The acoustic perturbation wave is known to freely cross the boundary layer surface, and this is the key to the solution.

As is seen from equation (5.3), the expression for the velocity u differs from the formulas for pressure $p$ and density ρ by an additional convective term.

Owing to the difference, the velocity in the wave packet changes its sign a little bit later than the pressure and density do. As a result, the wave carries some quantity of the fluid along with the frozen-in vorticity, and transfers it from the boundary layer to the potential part of the flow.

Interaction of the frozen-in vorticity with the potential flow of viscous fluid is determined by equation (2.19). The vortex as a kind of perturbation induces a circular motion of the fluid's adjacent particles of the potential part of the flow. The angular Ω and linear u velocities of the fluid particles in two-dimensional flow are expressed as follows [43]:

$$\left.\begin{aligned} \Omega &= \frac{\Gamma}{8\pi \nu t}\exp\left(-r^2/4\nu t\right), \\ \mathrm{u} &= \frac{\Gamma}{2\pi r}\left(1-\exp\left(-r^2/4\nu t\right)\right). \end{aligned}\right\} \tag{5.4}$$

The vortical motion transferred to the flow potential part by a perturbation wave creates therein some induced vorticity. The transferred vortex and induced vorticity are gradually smoothed and damped in the potential flow under the action of the viscosity friction represented in equations (5.4) due to the kinematic viscosity factor ν. This process generates potential flow perturbations comprising a broad range of the velocity u amplitudes directed in every ways.

The vortices transferred by a sequence of perturbation waves create a motion that appears as a disordered oscillation field of the fluid velocity. These oscillations correspond to the interpretation of turbulent fluctuations offered in [11]. Thus the wave conception of turbulent oscillations, or fluctuations as commonly said, implies two formation ways of the oscillations.

One important difference between the two ways, following from the expression for the velocity u in equation (5.3), should be noted here. The first item in the expression causes the formation of velocity oscillations according to the first mechanism. The second item is responsible for the oscillations formed amenably to the second mechanism. The comparison shows that the first term in equation (5.3) is inversely proportional to the distance covered by the perturbation wave, $r^{-1}$, whereas the second one is inversely proportional to the distance squared, $r^{-2}$.

Therefore, when $r \gg a$, the velocity oscillations are mainly determined by the first item, so that the velocity distribution within the wave packet becomes similar to the pressure and density distributions. The waves and oscillations related to the first mechanism exist for a relatively long time and cover rather a large distance. Since this process has not been analyzed in the literature so far, we will consider it in further detail.

## 5.2 Perturbation waves in flow

We have just seen that the wave front of a wave packet creates a spherical configuration in an infinite space filled with some medium at rest. When the wave propagates in a medium moving in a restricted space, e.g., in a channel, the wave front is distorted. The distortions arise, first of all, due to wave reflections from the channel walls, and secondly owing to the intricate velocity profile in the flow.

Let us consider a steady flow in a channel formed by two planes, where the flow velocity profile is of the form $U(z) = U_0[4z(1-z)]^{1/m}$. A perturbation is assumed to arise on the lower plane, its center being located at some point with random coordinates $\boldsymbol{rp}$ $(xp, yp, zp)$. The perturbations are observed at $x_0$, $y_0$, $z_0$, as shown in Fig. 5.1. The image of the velocity profile $U(z)$ is seen on the left from the channel scheme.

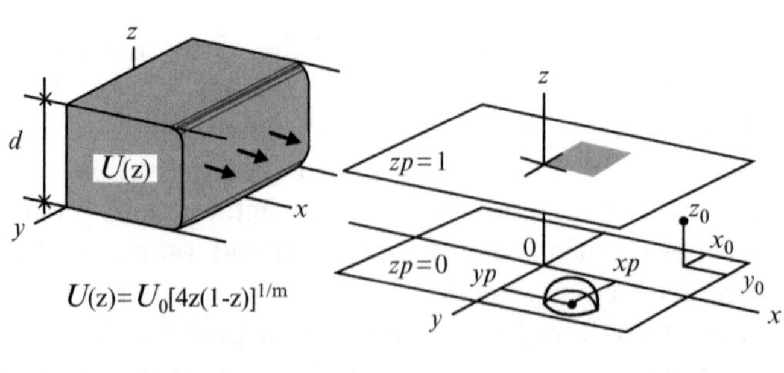


**Fig. 5.1.** A perturbation originated on a wall of a flat channel

The wave packet profile in the vertical plane is depicted in Fig. 5.2. Hear the wave propagation relative to the onset point $\boldsymbol{rp}(xp,0,0)$ would be described in the upper half plane (without reflections) by the radius-vector $\boldsymbol{\rho}(\xi,\eta,\zeta)$. The perturbation wave would pass through the observation point $\boldsymbol{r}_0(x,0,z)$ only one time in this case. When the wave packet propagates in the channel, it passes through this point many times. After multiple reflections, the packet creates at this point gradually fading oscillations, as the lower graph in Fig. 5.2 illustrates.

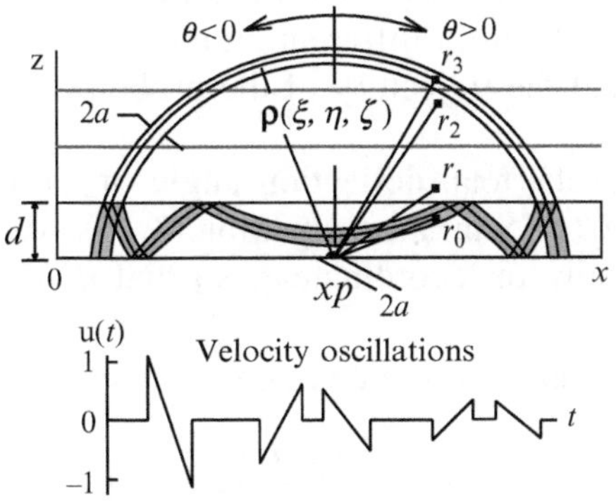


**Fig. 5.2.** The wave packet profile in a flat channel and velocity oscillations at the observation point $r_0$

The effect of the flow velocity profile $\boldsymbol{U}(\boldsymbol{r})$ on the character of the wave propagation can be determined using a known expression for the wave packet group velocity, $\boldsymbol{V} = (\partial\omega/\partial\boldsymbol{k})$, where $\boldsymbol{k}$ denotes the wave vector, and $\omega(\boldsymbol{k})$ is the dispersion law. A simple dispersion relation may be applied to the fluid moving at some moderate velocity $\boldsymbol{U}$ $(U \ll c)$:

$$\omega = ck + \boldsymbol{Uk}. \tag{5.5}$$

When the perturbation wave propagates in a flow, and the fluid velocity has some intricate profile, the wave front is distorted. In this case the path of some probe point and the shape of the wave front as a whole should depend not on the observation point distance $r$, but on the probe point trajectory length [44, 45]. Then, as the relations (5.3) suggest, the pressure $p$, density ρ, and velocity u in the wave packet, $|\ell - ct| \leq a$, vary along the trajectory at $\ell \gg a$ according to the relations

$$\varrho,\ p,\ \mathrm{u} \sim f\left(\frac{\ell - ct}{a}\right)\frac{1}{\ell}, \tag{5.6}$$

where the function $f(R/a)$ means the density or pressure distribution in the primary perturbation.

Now we need to define the trajectory of the probe point. Let $\boldsymbol{s}$ be a unit vector directed along the tangent to the trajectory at the probe point, which is determined by the equation ([3], §68]):

$$\frac{d\boldsymbol{s}}{d\ell}=\frac{1}{c}\left[\mathrm{rot}\,\boldsymbol{U},\,\boldsymbol{s}\right]. \tag{5.7}$$

When the fluid is at rest, $\boldsymbol{U} = 0$, the wave probe point moves at velocity $\boldsymbol{V} = c\boldsymbol{k}/k$. Accordingly, this point motion direction is $\boldsymbol{s} = \boldsymbol{\rho}/\rho$, and the chosen point of the wave moves along a straight line. If the fluid moves at constant velocity, $\boldsymbol{U}$ = const, then the point velocity is $\boldsymbol{V} = c\boldsymbol{k}/k + \boldsymbol{U}$. In this case the flow causes the wave to drift as a whole, and the vector $\boldsymbol{s}$ does not change its orientation, so that the wave keeps its spherical symmetry. Quite a different situation is observed near the wall, where $\boldsymbol{U}\neq$const, and the trajectory of the probe point curves, as it follows from equation (5.7).

Let us estimate the total deflection angle of the point trajectory in the flow velocity field $U(z)$. Similar to equation (3.15), we assume that the flow velocity depends only on coordinate $z$, so that $dU/dz > 0$ and $U = 0$, when $z = 0$.

Equation (5.7) can be rewritten in components of the vector $\boldsymbol{s}(s_x, s_y, s_z)$:

$$\left.\begin{aligned} \frac{ds_x}{d\ell} &= \frac{s_z}{c}\frac{dU}{dz}, \\ \frac{ds_y}{d\ell} &= 0, \\ s_z &= \sqrt{1-s_x^2+s_y^2}\,. \end{aligned}\right\} \tag{5.8}$$

Since $s_z = dz/d\ell$ by definition, integration of equations (5.8) gives:

$$\left.\begin{aligned} s_x &= s_{x0} + M(z), \\ s_y &= s_{y0}, \\ s_z &= \sqrt{1-s_x^2-s_y^2}, \end{aligned}\right\} \tag{5.9}$$

where $\boldsymbol{s}_0(s_{x0}, s_{y0}, s_{z0})$ are initial values of the vector $\boldsymbol{s}$ and the components, and $M(z) = U(z)/c$ is the Mach number of the flow at the ordinate $z$.

Introducing polar $\vartheta$ and azimuthal $\varphi$ angles of the vector $\boldsymbol{s}$, we obtain the equations for variable orientation $\{\vartheta, \varphi\}$ of the probe point trajectory in the form of a dependence on the trajectory initial direction $\{\vartheta_0, \varphi_0\}$ and coordinate $z$:

$$\left.\begin{aligned} \sin\vartheta\,\cos\varphi &= \sin\vartheta_0\cos\varphi_0 + M(z), \\ \sin\vartheta\,\sin\varphi &= \sin\vartheta_0\sin\varphi_0\,. \end{aligned}\right\} \tag{5.10}$$

The variable trajectory orientation can be obtained from equations (5.10) in the form of the following two expressions [46, 47]:

$$\left.\begin{aligned} \sin\vartheta &= \pm\sqrt{\left(\sin\vartheta_0\cos\varphi_0 + M(z)\right)^2 + \sin^2\vartheta_0\sin^2\varphi_0}\,, \\ \sin\varphi &= \frac{\sin\vartheta_0\sin\varphi_0}{\sqrt{\left(\sin\vartheta_0\cos\varphi_0 + M(z)\right)^2 + \sin^2\vartheta_0\sin^2\varphi_0}}. \end{aligned}\right\} \tag{5.11}$$

Our knowledge of the vector $\boldsymbol{s}$ gives us an opportunity to find the trajectory $\ell(\boldsymbol{r})$ of the probe point moving in the field of the variable flow velocity. First, we find the trajectory $\ell(\boldsymbol{\rho})$ in relation to the centre of the wave origin (without reflections), assuming that the point starts moving at the path length and coordinates $\ell = \xi = \eta = \zeta = 0$, when $t = 0$. The point's subsequent location is defined by dimensionless values of the coordinates $\boldsymbol{\rho}(\xi,\eta,\zeta)$ scaled by the channel height $d$.

To be definite, we discuss the case when the perturbation arises on the bottom wall. According to equalities (5.10), the probe point coordinates of the trajectory depend on the path length $\ell$ and are determined by the following parametric differential equations [45]:

$$\left.\begin{aligned} \frac{d\xi}{d\ell} &= \sin\vartheta_0\cos\varphi_0 + M(z); \\ \frac{d\eta}{d\ell} &= \sin\vartheta_0\sin\varphi_0; \\ \frac{d\zeta}{d\ell} &= \sqrt{1 - \left(\sin\vartheta_0\cos\varphi_0 + M(z)\right)^2} \end{aligned}\right\} \tag{5.12}$$

at the initial conditions

$$\xi(0) = \eta(0) = \zeta(0) = 0. \tag{5.12a}$$

## 5.3 Distortion of the wave packet in channels

Since the real wave propagates in a channel, say, formed by two plane walls, we need to use new coordinates related to the channel with the origin at $x = 0$, $y = 0$, $z = 0$. As before, the perturbation wave arises on one of the walls in the centre $\boldsymbol{rp}(xp,yp,zp)$ at some instant $tp = \tau c/d$ in relation to the fixed time $t = 0$. This time the radius-vector $\boldsymbol{\rho}(\xi,\eta,\zeta)$ indicates location of the probe point relative to the center $\boldsymbol{rp}(xp,yp,zp)$ in the new coordinate system. Now the wave reflections from the walls should be taken into account in describing the wave configuration in addition to equations (5.12).

The quantity $n$ of the wave reflections in the channel under consideration depends on the probe point ordinate $\zeta$. Looking at the diagram shown in Fig. 5.2, the quantity may be expressed in the form of $n = \mathrm{In}(\zeta)$, where the sign In means the operation of taking the whole part of the value $\zeta$. The quantity $n$ is determined by the formula

$$n = \mathrm{In}\left[\int_0^{\ell - tp} \sqrt{1 - \left(\sin\vartheta_0 \cos\varphi_0 + M(z)\right)^2}\, d\ell\right]. \tag{5.13}$$

Taking into account the place $\boldsymbol{rp}$ and time $tp$ of the wave origin results in transformation of equations (5.12) to the following form [36]:

$$\left.\begin{aligned} x &= xp + \int_0^{\ell - tp} \left(\sin\vartheta_0 \cos\varphi_0 + M(z)\right) d\ell; \\ y &= yp + \sin\vartheta_0 \sin\varphi_0 \, (\ell - tp); \\ z &= zp + \int_0^{\ell - tp} \sqrt{1 - \left(\sin\vartheta_0 \cos\varphi_0 + M(z)\right)^2}\, d\ell - \\ &\quad - 2\,\mathrm{In}\left[\frac{n+1}{2}(-1)^{n+zp}\right]. \end{aligned}\right\} \tag{5.14}$$

The effect of the flow velocity gradient on properties of the probe point trajectory depends on the angles $\vartheta$ and $\varphi$ with other parameters being equal. To illustrate the effect, the trajectory is considered for $n=0$ in the plane $\vartheta=0$, $\varphi=0$, where the influence is seen to be the best of all planes. Under the given conditions the angular relations (5.10) are reduced to the expression

$$\sin\vartheta = \sin\vartheta_0 + M_0\,[4z(1-z)]^{1/m}, \tag{5.15}$$

and the corresponding equations (5.12) take the form

$$\left.\begin{aligned} \frac{d\xi}{d\ell} &= \sin\vartheta_0 + M(z); \\ \frac{d\zeta}{d\ell} &= \sqrt{\cos^2\vartheta_0 - 2\,M(z)\,\sin\vartheta_0 - M^2(z)}\,. \end{aligned}\right\} \tag{5.16}$$

The equality (5.15) suggests some valuable conclusions about properties of the probe point trajectory. If the trajectory's initial direction belongs to the region of non-positive angles $\vartheta_0 \le 0$, in Fig. 5.2 it corresponds to the left quadrant $x \le xp$, then the probe point may move off from the wall for any arbitrary distance $z$, certainly within the channel height $d$.

And vice versa, in the region of positive angles $\vartheta_0 > 0$ and $x > xp$, right quadrant in the Fig. 5.2, there exist such quantities of the parameters $M_0$, $m$

and of the angle $\vartheta_0 \geq \vartheta^*$, when the distance of the wave probe point from the wall does not exceed some upper limit $z=z^*$. This limit corresponds to the angle $\vartheta=\pi/2$.

In other words, the set of these three parameters, $M_0$, $m$ and $\vartheta_0$, determines the limit value of $z^*$, if the limit generally exists for the set. The quantity $z^*$ makes sense of the trajectory singularity. It is clear when the inverse function is treated.

Under this condition, quantity $z$ as an explicit function of $M_0$, $m$ and $\vartheta_0$ can be deduced from formula (5.15):

$$z = \frac{1}{2}\left(1 \mp \sqrt{1 - \left(\frac{1-\sin\vartheta}{M_0}\right)^m}\right). \tag{5.17}$$

The sign 'minus' before the square root corresponds to the lower wall, and the sign 'plus' relates to the upper wall. As we have agreed, the distance z possesses the limit value $z^*$ when $\vartheta=\pi/2$ and $\vartheta_0=\vartheta^*$.

The expression (5.17) predicts an interesting dependence of the ordinate $z^*$ on the initial angle $\vartheta_0$. The ordinate is a complex number within the angle interval $\vartheta_0 < \vartheta^*$, so that the trajectory of the probe point is limited only by the walls of the channel. At the same time, the ordinate in the angle interval $\vartheta_0 \geq \vartheta^*$ is a real number. The function $z^*(\vartheta_0)$ in the interval has a maximum, while the function $\vartheta_0(z)$ discontinues at the point $z^*$. The function $z^*(\vartheta_0)$ on the right-hand side of point $z^*$, in the interval $\vartheta_0 > \vartheta^*$, has the same maximum. Hence the trajectory describes some kind of an arc-wise curve to return again to the wall.

The function $z^*(\vartheta_0)$ depends, as seen from expression (5.17), not only on the initial angle $\vartheta_0$, but also on the parameter $m$ and velocity $M_0$. The graphs of the function are depicted in Fig. 5.3 as the surfaces $z^*(m, M_0, \vartheta_0)$ for the parameters $m$ = 1, 5 and 10.

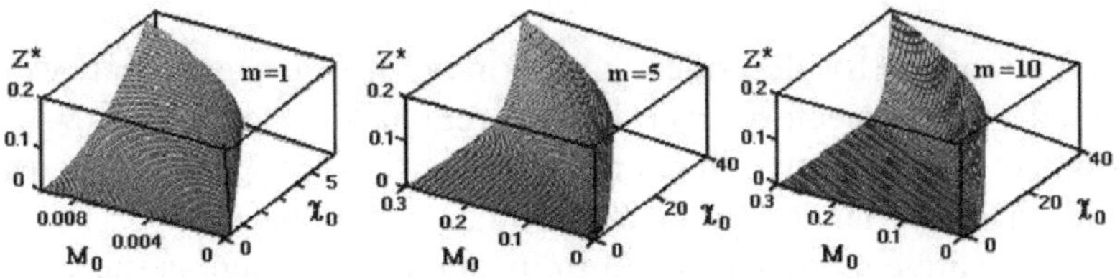


**Fig. 5.3.** Function $z^*(m, M_0, \vartheta_0)$ for $m$ = 1, 5, 10

The graphs plotted in Fig. 5.3 have been computed for air in a channel with height $d$ = 1 cm. The polar angle $\vartheta_0$ is replaced in the graphs with its trajectory inclination angle relative to the wall, $\chi_0 = \pi/2 - \vartheta_0$. The

Reynolds number in this instance reaches its critical value at the velocity $M_0 \approx 0.01$. Wherefore the function $z^*(m, M_0, \chi_0)$ is plotted in the velocity interval $M_0 \in [0,\ 0.01]$ for the flow's laminar mode, when $m = 1$. As to the flow's turbulent mode, when $m = 5$ and $m = 10$, the interval comes to $M_0 \in [0.01,\ 0.3]$, thus the fluid can still be considered as non-compressible.

The surface $z^*(m, M_0, \chi_0)$ divides the perturbation wave into two sectors. In the sector of small angles $\chi_0$ ($\vartheta_0 \to \pi/2$), the parameter $z^*$ being a real value defines the trajectory ordinate maximum and its subsequent motion to the wall. In the sector of large angles $\chi_0$ ($\vartheta_0 \in [0, \vartheta^*]$), the parameter $z^*$ is a complex value. Therefore there exists no backtracking point at all, so that the trajectory leaves the field of the flow velocity gradient to make its way into the potential part of the flow.

The shape of the surface $z^*(m, M_0, \chi_0)$ and size of the angle $\vartheta^*$ (or $\chi^*$) depend on the flow mode. Indeed, when $m = 1$, the probe point backward motion does occur, if the trajectory belongs to a small sector of the wave adjoining a wall, $\vartheta_0 > 85°$. Hence, only a very small portion of the wave energy remains within the boundary layer to create perturbations.

When the parameter $m$ increases, this sector widens, so that if $m = 10$, the probe point backward motion occurs in the large angle interval $\vartheta_0 > 50°$. The larger portion of the wave energy is used for creation of perturbations within the boundary layer. Besides, while the parameter $m$ grows, the function $z^*(\vartheta_0)$ increases more and more steeply.

To study special features of the trajectory, it is better to deal with its dependence in the explicit form, $z(x)$. For that purpose we take advantage of the parametric equations (5.16). Let the channel be constituted of two parallel planes, and the perturbation arise at $\boldsymbol{rp} = tp = 0$. Simple transformations of the equations lead, for these requirements, to the following equation for the trajectory:

$$\frac{dx}{dz} = \frac{\sin^2 \vartheta_0 + M_0 (4z(1-z))^{1/m}}{\sqrt{1 - \left(\sin \vartheta_0 + M_0 (4z(1-z))^{1/m}\right)^2}} \tag{5.18}$$

Equation (5.18) describes both kinds of trajectory that we have discussed. The trajectory $z(x)$ in the region of the angles $\vartheta_0 < \vartheta^*$ is determined by the usual numerical integration of (5.18), whereas the integration in the angle region $\vartheta_0 > \vartheta^*$ has some specificity. The issue is that the derivative $dx/dz$ tends to infinity and changes sign in the vicinity of the point $z^*$. Therefore, it is necessary, first, to define the position$x^*$ of the trajectory maximum $z^*$ relative to the wave centre $\boldsymbol{rp}$, and then to integrate separately at the left and at the right of this maximum.

The Program 5.1 is an example of the problem solution. The trajectory is calculated for the flow parameters $m = 5$ and $M_0 = 0.05$, with $\vartheta_0 = 75°$.

**Program 5.1** Computation of trajectory of the wave probe point

$m=5;\ M_0\approx 0.05;\ \vartheta_0=75°;$

$t_0=\pi\vartheta_0/180;\ \vartheta=\text{ArcSin}[\text{Sin}[\pi\vartheta_0/180]+M_0(4z(1-z))^{1/m};$

$$z^*=\frac{1}{2}\left(1-\sqrt{1-\left(\frac{1-\text{Sin}[t_0]}{M_0}\right)^m}\right);$$

$x^*=\text{NIntegrate}\left[\text{Tan}\left[\text{ArcSin}[t_0]+M_0(4z(1-z))^{1/m}\right],\ \{z,0,z1\}\right];$

$z1=(1-10^{-8})z^*;\ x1=(1+10^{-4})x^*;$

$xL=\text{NIntegrate}\left[\text{Tan}\left[\text{ArcSin}[t_0]+M_0(4z(1-z))^{1/m}\right],\{z,0,i(z1/2000)\}\right];$

$xR=\text{NIntegrate}\left[\text{Tan}\left[\text{ArcSin}[t_0]+M_0(4z(1-z))^{1/m}\right],\{z,z1,z1(1-(i/2000)\}\right];$

$xLi=\text{Table}\left[\{xL,\ i(z1/2000)\},\{i,0,2000\}\right];$

$xRi=\text{Table}\left[\{xR+x1,\ z1(1-(i/2000)\},\{i,\ 2000\}\right];$

$g1=\text{ListPlot}[xLi,\text{PlotJoined}\to\text{True},\ \text{Frame}\to\text{True},$
$\quad\text{PlotRange}\to\{\{0,2.1x^*\},\ \{0,1.1z^*\}\},$
$\quad\text{FrameLabel}\to\{\text{“}x\text{”},\text{”}z\text{”}\},\ \text{DisplayFunction}\to\text{Identity}];$

$g2=\text{ListPlot}[xRi,\text{PlotJoined}\to\text{True},\ \text{Frame}\to\text{True},$
$\quad\text{PlotRange}\to\{\{0,2.1x^*\},\ \{0,1.1z^*\}\},$
$\quad\text{FrameLabel}\to\{\text{“}x\text{”},\text{”}z\text{”}\},\ \text{DisplayFunction}\to\text{Identity}];$

$gzx=\text{Show}[\{g1,g2\},\ \text{DisplayFunction}\to\$\text{DisplayFunction}];$

As usual, the first two lines of the program contain some set of initial conditions and formulas for definition of the parameters in practical use. The next three lines include computation of the ordinate z*and position x* of the maximum, and then the coordinates of z1, x1, the nearest to z*, x*, where the calculation can be fulfilled.

After that the dependences of abscissa xL on ordinate z along the trajectory left part, x<x*, and the dependence of abscissa xR on ordinate z along the trajectory right part, x>x*, are determined. These data are summarized in tables xLi and xRi. In the end, the procedures g1, g2 prepare the table data for visual representation, and the operation gzx plots the dependence graph.

The program provides a computation for one complete cycle of the probe point motion. The cycle includes the phases of the point moving off from the wall and its backward motion. However, the trajectory can be computed for any length according to the amplitude's prescribed attenuation. In particular, the trajectory may overstep the limits of one cycle.

## 5.4 The wave packet in the boundary layer

To gain a more comprehensive understanding of the effect of the flow velocity gradient on certain features of the disturbances, we consider the wave shape in the flow of laminar and turbulent modes. The shape variation can be ascertained by comparison of the trajectories of some probe points. We choose, for demonstration, trajectories at seven initial orientations $\vartheta_0$ in the flow of the velocity profile (3.15), $M(z) = M_0[4z(1-z)]^{1/m}$, at two couples of the parameters, $m = 1$, $M_0 = 0.005$ (laminar mode) and $m = 5$, $M_0 = 0.05$ (turbulent mode). Also, some useful information can be obtained by comparison of the computed trajectories with the imaginary ones that might occur in the hypothetical flow with $m = \infty$. As has been said in comments to equation (5.7), the trajectories in the latter case do not bend but rather appear as straight lines.

The problem in all enumerated alternatives is solved by means of Program 5.1. Outcomes of the solutions for the chosen parameters are represented by diagrams in Fig. 5.4. The wave front in the laminar flow is on the left, and the front in the turbulent flow is on the right. The wave fronts on both graphs correspond to the time instant $t = 1$ (in dimensionless units). Since the flow velocity is small in comparison with the wave velocity, the wave front looks like a half-round with radius $\ell = 1$. The radial lines represent the trajectories for the initial angles $\vartheta_0 = 0, \pm25, \pm50, \pm75°$ for all the mentioned values of the parameters $M_0$ and $m$, including $m = \infty$.

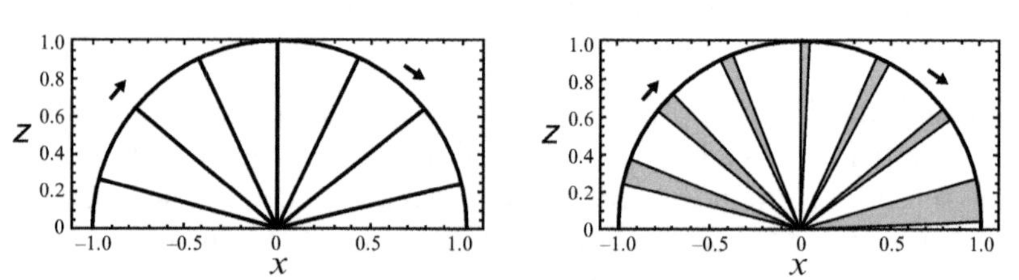


$m = 1$ and $m = \infty$, $M_0 = 0.005$ $m = 5$ and $m = \infty$, $M_0 = 0.05$

**Fig. 5.4.** Trajectories at various flow velocity, parameter $m$ and angle $\vartheta_0$

As follows from the graphs in the figure, when the perturbation wave propagates in the laminar flow, $M(z) = M_0[4z(1-z)]$, the trajectory curvature is too insignificant to be detected, whatever the angle $\vartheta_0$ might be. Also, the trajectories at $m = 1$ and $m = \infty$ practically coincide and appear to be straight lines, at least for the picture scale. As to the turbulent flow, when the flow velocity is $M(z) = M_0[4z(1-z)]^{1/5}$, every trajectory deflects from the straight line ($m = \infty$) clockwise. The area of the deflection is shaded in the figure. The amount of deflection depends on the initial angle $\vartheta_0$, and it is most considerable at $\vartheta_0 = 75°$ of all the chosen angles.

While $|\vartheta_0|<|\vartheta^*|$, the trajectory deflection from the initial direction is small, and it can be neglected similarly to the laminar flow. If $\vartheta_0 \Rightarrow -\pi/2$, the trajectory moves away from the wall. But when $\vartheta_0 > \vartheta^*$ and $\vartheta_0 \Rightarrow \pi/2$, the deflection becomes more and more essential, and the effect must necessarily be taken into account. The effect's fundamental importance is illustrated by two trajectories at different values of the parameter $m$, with other parameters being equal, namely, the initial angle $\vartheta_0 = 83°$ and flow velocity $M_0 = 0.05$. The trajectories are shown in Fig. 5.5 [34, 47].

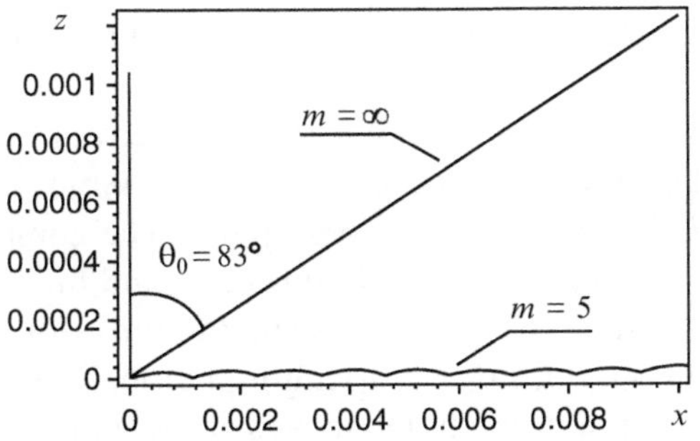


**Fig. 5.5.** The trajectories at $m = \infty$ and $m = 5$ with flow parameters $\vartheta_0 = 83°$, $M_0 = 0.05$

It is astonishing to see such unlike curves resulted under nearly identical conditions. However it is necessary to note that the trajectory calculation for the angle $\vartheta_0 = 83°$ has been fulfilled by means of Program 5.1 for a formally the case, when the trajectory corresponds to an infinitesimal angular aperture. Meanwhile such a trajectory cannot exist in reality, in particular when the wave packet thickness $2a$ exceeds considerably the layer depth $z^*$ typical for initial angle $\vartheta_0 = 83°$.

Therefore we will continue consideration for the sake of the principal importance of trajectory behavior, but when the angle $\vartheta_0$ is large enough though does not belong in close proximity to angle $\pi/2$. Thus we trace the deflection of the probe point trajectory from the straight line in the region of initial angles $\vartheta_0 = 70, 73, 75, 76, 77°$. In order to simplify the graphic picture, we consider only the narrow flow layer, $z \in [0, 0.1]$, and the trajectories by the length $\ell \approx 1.2$ each at the primary perturbation radius being $a = 0.04$.

The wave amplitude of such perturbation within the distance $\ell \approx 1.2$ decreases approximately 30 times. This means that consideration of the amplitude's subsequent lessening may be omitted, having in view the phenomenon illustration. Integration of equation (5.18) for the listed initial angles $\vartheta_0$ gives the trajectories presented in Fig. 5.6 [35, 36].

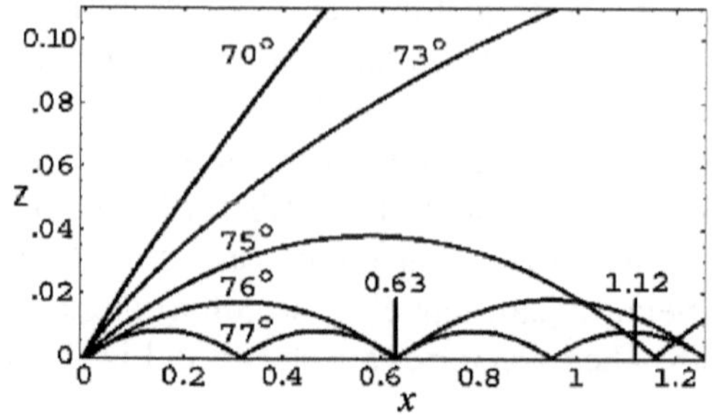


**Fig. 5.6.** Trajectories of the probe point near the wall

The trajectory at the initial angle $\vartheta_0=70°$ leaves the layer $z\in[0, 0.1]$, that maintains the motion direction almost invariable. But when the angle runs up to $\vartheta_0 = 73°$, it already describes an arc. The complete cycle is of the length $\ell \approx 4.5$ and altitude $z^* \approx 0.15$ in this case, and the trajectory maximum can not be seen on Fig. 5.6, for it is beyond the limits of the graph.

The perturbation within such a length practically fades away (accurate within ~3%) before the probe point arrives at the end of the complete cycle. There remain within the selected layer only the trajectories with initial angles $\vartheta_0$ = 75, 76, 77°. While the path length covered by the probe point is $\ell \approx 1.2$, each of the trajectories with these initial angles has time to describe more than one cycle, interlacing with one another. Thus, for example, in the cross-section $x$ = 1.16 we observe trajectories with the sequence of the angles 75-77-76°, instead of the angle sequence 77-76-75°, as it might be expected.

Of course, the angle $\vartheta_0$ varies continuously within the perturbation wave from –90° to +90°. So that when $\vartheta_0 \geq 75°$, a multitude of the elementary trajectories intermix completely within the layer $z^*(75°) \approx 0.04$. At all that, there appear some spots on the wall surface, in which the wave reflection positions simultaneously coincide with each other. For instance, such a spot is observed around the abscissa $x = 0.63$ on the graph in Fig. 5.6.

The trajectory curvature's strong dependence on the angle $\vartheta_0$ in the turbulent flow has a simple explanation. The deflection is determined by the path length $\ell$ of the wave probe point, which passes through the flow field of the heavy velocity gradient. The gradient in the fluid laminar motion is small throughout all the flow cross-section, as seen in Fig. 3.1. Hence the trajectory curvature and the deflection from the straight line seem insignificant in the laminar flow.

As to the turbulent flow, it is quite another matter. The velocity gradient in the main potential stream is small too. But the gradient is reinforced harshly, as far as the ordinate $z$ tends to the wall. And a line that would go near the wall in the layer by thickness δ in the direction $\vartheta_0$ is of the length $\ell_0 \approx \delta/\cos\vartheta_0$ which rapidly increases when the angle $\vartheta_0$ grows. Curvature of the line going through the flow gradient field strengthens this tendency much more still, i.e., the distance $\ell$ lengthens much quicker as compared with $\ell_0$. Therefore, the trajectory curvature displays strong dependence on the initial angle $\vartheta_0$. The strengthening within a narrow range of the initial angles $\vartheta_0$ near the wall results in a rather interesting phenomenon.

The trajectories at this narrow range of the angles $\vartheta_0$ are concentrated at some certain distance $z$. When the angle diminishes slightly, the trajectories leave the wall vicinity to propagate freely in the main potential flow [33, 34]. This abrupt change of the trajectory behavior is demonstrated by the graph in Fig. 5.7. Here solutions of the equations (5.18) are represented for the interval $\vartheta_0 \in [72, 73°]$ with the argument step $\Delta\vartheta_0 = 0.1°$.

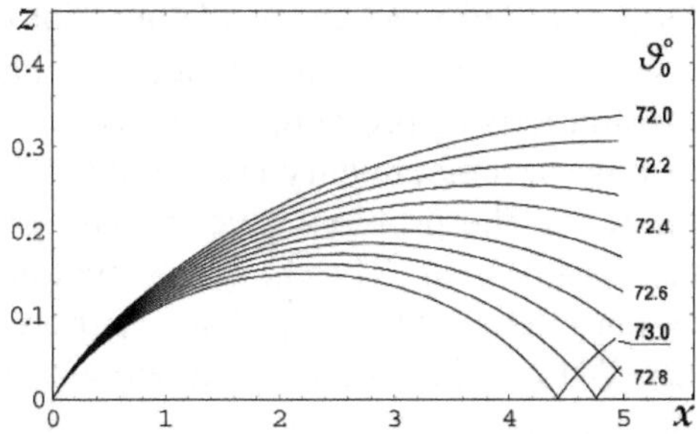


**Fig. 5.7.** Trajectories near the wall for a set of the angle $\vartheta_0 \in [72, 73°]$

Here the length of the trajectories amounts to $\ell = 5$ which means 100-fold decrease of the wave amplitude.

The trajectory curvature dependence on the initial angle $\vartheta_0$ has one more important consequence. There are some spots on the wall surface where several wave packets of thickness $2a$ are simultaneously reflected from the wall. This implies the pressure and density local augmentation in the layer by height $z \le a$. A wave propagation in a medium with a variable density is known to be a nonlinear process ([3], §66). Nonlinearity is one of the causes of wave energy dissipation. The nonlinear interaction in the case under consideration occurs in the same layer where the trajectories interlace. Hence, there appears an additional opportunity to estimate the characteristic dimension of the primary perturbation $a$ [34, 35, 47]:

$$a = z^*(\vartheta^*). \tag{5.19}$$

Recall that, in a thin layer of the flow, $z \ll z^*$, immediately adjoining the wall surface, the fluid moves under the influence of the molecular viscous friction. The velocity is small throughout this thin layer, varying from zero at the wall surface to values of the order of $U \sim (\sigma/\varrho)^{1/2}$ at $z \sim$

$\nu(\varrho/\sigma)^{1/2}$. Herein the velocity distribution is a linear function of the ordinate $z$ (2.13).

Researchers used to call this thin layer as the laminar sublayer because of its special characteristics. Though broad-minded, L.D. Landau ([3], §42) preferred the term "viscous sublayer". He emphasized that "the flow in the viscous sublayer is turbulent." "In this respect the name 'laminar sublayer' still sometimes used is unsuitable. The resemblance to laminar flow lies only in the fact that the mean velocity is distributed according to the same law as the true velocity would be for a laminar flow under the same conditions. The fluctuating flow in the viscous sublayer has some peculiar features that have not yet been given an adequate theoretical explanation."

However, the wave conception of turbulent oscillations gives an adequate explanation for the turbulence nature in the viscous sublayer. Indeed, the velocity oscillations in the wave packet imply some energy flux $c\varrho u^2$ ([3], §65). As the velocity on the wall turns into zero, the velocity oscillations vanish at the wave packet reflection, and their energy is transmitted to the oscillations of pressure and density. That is why there is no sharp boundary between the boundary layer and the rest of the flow.

The wave approach to the problem gives us an opportunity to size up the viscous sublayer thickness $\delta_L$. The thickness obviously depends on the distribution of the parameters in the primary perturbation, i.e., on the function $f[(\ell - ct)/a]$. With various kinds of functions, the value of $\delta_L$ may be estimated as

$$\delta_L \sim (0.1 - 0.2)\ a. \tag{5.20}$$

We should make an essential remark. As it follows from all said, the trajectory intersections in the layer $z < z^*$ form some new peculiar spatio-temporal field of the parameter oscillations. To understand the field's true sense, we should have in view, that the equation (5.7) used in the above consideration has been deduced in the approximation of geometrical acoustics (see [3], §67), when the trajectory is actually the sound ray ([3], §68). But the intersections of the sound rays mean that geometrical acoustics is inapplicable in these conditions, for the diffraction effects. We only may state, that the regular shape of the wave with a spherical front terminates to exist as a single whole.

The facts concerning the wave propagation mode in the layer by thickness $z^*$ emphasize sharp distinction between the main flow and the one in the layer. The surface at the level of the ordinate $z^*$ is formed by the trajectories along which the perturbations remain still of significant amplitudes, at least within one cycle. Also the distinction implies the strong gradient field of the flow longitudinal velocity and vorticity in the layer, which is the sign of the boundary layer. As the graphs in Figs. 5.6 and 5.7 ($m = 5$, $M_0 = 0.05$, $\vartheta^* = 75°$) show, the boundary layer definition corresponds to the condition [36]

$$\delta \approx z^*(\vartheta^*) = 0.04. \tag{5.21}$$

The analysis carried out above has allowed us to reveal some physical processes in the flow field of the strong velocity gradient, owing to which the boundary layer and viscous sublayer are formed. Also, the analysis has allowed us to estimate thickness of the boundary layer and sublayer, as well as the primary perturbation size.

It goes without saying, that the parameter primary perturbations determine the boundary layer structure. But the perturbations exert influence upon the structure of the main flow as well. The main flow structure is investigated in the next section.

# 6
# Spatio-temporal field of perturbations in channels

## 6.1 Computing technique of wave configuration in channels

According to the estimations given in equation (4.17), formation process of primary perturbations of flow parameters runs within the small volume V practically in a trice, as compared with other events in the flow. These primary perturbations propagate throughout the flow as acoustic wave packets damping gradually (5.3).

The structure of the flow perturbations in channels is defined by propagation of the primary wave packets, direct and reflected from the channel walls, and by features of the sequence in which the perturbations arise on the wall surfaces. In its turn, this structure representing the propagating wave distribution produces the spatio-temporal field of the parameter oscillations.

The spatio-temporal field in a channel depends besides on the configuration of the channel, which transforms the shapes of the waves after their multiple reflections from the walls. The shape of each wave front can be computed by means of the relations (5.13), (5.14). These equations are strongly simplified, when the main part of the flow is considered, where the flow velocity gradient can be neglected [45]:

$$\left.\begin{aligned} &x=(t-tp)\sin\vartheta\cos\varphi+xp,\\ &y=(t-tp)\sin\vartheta\sin\varphi+yp,\\ &z=\left((t-tp)\cos\vartheta-2\,\mathrm{In}\left[\frac{n+1}{2}\right]\right)(-1)^{n+zp},\\ &x^2+y^2+z^2>a^2,\end{aligned}\right\} \tag{6.1}$$

$$n=\mathrm{In}[\zeta]=\mathrm{In}[(t-tp)\cos\vartheta]. \tag{6.2}$$

Let us appeal to Fig. 5.2 and remember, that the ordinate $\zeta_0$ of the perturbation front at some observation point $\boldsymbol{r}_0(x_0,y_0,z_0)$ is simply equal to this point ordinate, $\zeta_0 = z_0$. When the perturbation passes the observation



L.N. Pyatnitsky, *Turbulence Nature and the Inverse Problem*, Fluid Mechanics and Its Applications 89, 73–84.

point after $n$ reflections, the wave front ordinate $\zeta_n$ coincides with the ordinate $z_n$ of the virtual image of the observation point, $\boldsymbol{r}_n(x_0, y_0, z_n)$.

The first virtual image is simply a mirror image of the observation point in the wall plane. Its ordinate $\zeta_1$ is to be expressed by the evident relation $\zeta_1 = 2d - \zeta_0$. The next reflections add to the imaginary ordinate some value which depends on whether the reflection number $n$ is even, or odd. The first variant corresponds to the quantity $2(d-\zeta_0)$, whereas in the second case it amounts to $2\zeta_0$. All this means, that

$$\left.\begin{aligned} \zeta_{n+1} &= \zeta_n + 2(d - \zeta_0), \text{ if } n \text{ is even,} \\ \zeta_{n+1} &= \zeta_n + 2\zeta_0, \text{ if } n \text{ is odd.} \end{aligned}\right\} \tag{6.3}$$

The diagram in Fig. 5.2 illustrates the simple rule for the first three virtual images, $r_1$, $r_2$, $r_3$ of the observation point $r_0$.

The convenient rule can be easily extended to the case of a channel in the form of a pipe, in particular with a polygonal cross-section [45]. The calculation pattern of the perturbation wave configuration in such a pipe works on the same principle as in the plane channel, when the relations (6.1)–(6.3) are employed. Though the wave reflections simultaneously from the pipe several sides need to take into account the imaginary points for every side.

To demonstrate the computing technique for determining the wave configuration in the pipe of a polygonal cross-section, we shall consider the pipe with a square section. First of all, it must be taken into account that the wave reflection quantity per a cycle trebles in the pipe of square section, and the ordinate of every imaginary point is defined not by one, but by two ordinates. These ordinates change according to formula (6.3) in which the length of the square side $d$ is implied as a scale, and the coordinate system is similar to the one of Figs. 5.1 and 5.2.

The perturbation wave configuration evidently does not depend on the abscissa $x$ of the perturbation origin site of the pipe. However the configuration does strongly depend on the wave's initial coordinates on the pipe walls, i.e., in the directions of the axes $y$ and $z$. We denote the initial coordinates on the walls parallel to the plane $yx$ at $z = 0$ through $yp0$, and through $yp1$ at $z = 1$, where $z$ is a dimensionless value. In similar way, the initial coordinates on the walls parallel to the plane $zx$ can be designated as $zp0$ ($y = 0$) and $zp1$ ($y = 1$). Then the coordinates $yp0$ and $zp0$ are evidently to vary arbitrary in the limits from 0 to 1, whereas coordinates $yp1$ and $zp1$ may only be equal to 0 or 1.

Now it is necessary to define the number of reflections for the wave moving from any one wall of the channel toward the others. We will distinguish the wave motion from the horizontal walls, when $z = 0$ or $z = 1$, and from vertical walls, when $y = 0$ or $y = 1$. To avoid misunderstanding henceforth, the wall notations are shown by the diagram in Fig. 6.1. Each of the mentioned directions implies wave reflections from three other walls, and subsequent reflection from the wall of the wave's start.

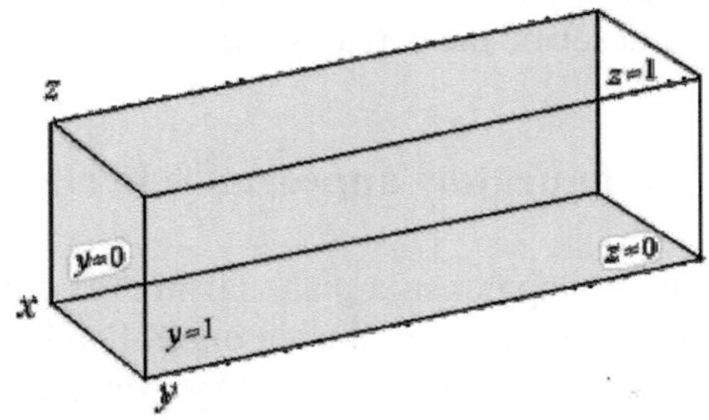


**Fig. 6.1.** Coordinate system concerning the channel in use

Let the reflection number for the wave starting on one of the horizontal walls at $z = 0$ (or $z = 1$) be named $ny0$, $ny1$, $nyz$, when correspondingly $y = 0$, $y = 1$, $z = 1$ (or $z = 0$). In the same manner, if the wave starts on one of the vertical walls at $y = 0$ (or $y = 1$), the reflection number from one of the walls at $z = 0$, $z = 1$, $y = 1$ (or $y = 1$) is designated as $nz0$, $nz1$, $nzy$, accordingly.

By analogy with formula (6.2), the quantity $n$ of the wave outgoing from one of two horizontal walls ($z = 0$ or $z = 1$) and reflected from one of the other walls is defined by equations (6.4a). In a similar manner, the quantity $n$ of the wave outgoing from one of two vertical walls ($y = 0$, or $y = 1$) and reflected from one of the other walls is defined by equation (6.4b) [45]:

Equations for the waves outgoing from horizontal walls:

$$\left.\begin{aligned} ny0 &= \mathrm{In}\left[(t-tpy)\sin(\vartheta-\pi/2)+yp0-1\right], \\ ny1 &= \mathrm{In}\left[(t-tpy)\sin(\vartheta-\pi/2)+yp0\right], \\ nyz &= \mathrm{In}\left[(t-tpy)\cos(\vartheta-\pi/2)+ypz\right]; \end{aligned}\right\} \tag{6.4a}$$

Equations for the waves outgoing from vertical walls:

$$\left.\begin{aligned} nz0 &= \mathrm{In}\left[(t-tpz)\cos\vartheta+zp0-1\right], \\ nz1 &= \mathrm{In}\left[(t-tpz)\cos\vartheta+zp0\right], \\ nzy &= \mathrm{In}\left[(t-tpz)\sin\vartheta+zpy\right]. \end{aligned}\right\} \tag{6.4b}$$

Parameter $t$ in a channel with a square cross-section is the current time (in dimensionless units), just the same as in the previous case of a two-dimensional flow. Parameters $tpy$ and $tpz$, relating to the horizontal and vertical walls, mean a time delay of a perturbation origin concerning the instant $t = 0$. These parameters may be of positive sign, real delay, or negative one, when the perturbation arises before the instant $t = 0$. The polar angle varies within interval $\vartheta \in [0, \pi]$. Three ordinates $\zeta$ appear for three reflections of the wave from three different walls of the pipe, instead of one ordinate $\zeta$, to describe the reflected wave configuration.

## 6.2 Wave front configuration appearance in channels

While a perturbation propagates in a plane channel, a current configuration of the wave reflected from one wall during one reflection cycle is specified by variables $x,y,z$ which are determined by relations (6.1). The wave configuration in a pipe depends on the reflections from three walls during one cycle, and three parameters are necessary for each variable of the set $x,y,z$. The formulas defining the parameters do not basically differ from the ones presented in equation (6.1). Therefore, we can save room and consider the example of the wave configuration in the flow plane $\varphi = \pi/2$, which comes through the abscissa $x = 0$.

The reflected wave configuration in the pipe under these conditions is specified by new variables $y0$, $y1$, $zy$ instead of $y$, and in much the same way by the variables $z0$, $z1$, $yz$ instead of $z$. These new variables can be evaluated by means of equations (6.5a) that replace the second equation in system (6.1), and by equations (6.5b) that replace the third equation in system (6.1).

$$\left.\begin{aligned} y0 &= \big((t-tpy)\sin(\vartheta-\pi/2)+ypy-2\,\mathrm{In}\left[ny0/2\right]\big)(-1)^{ny0},\\ y1 &= \big((t-tpy)\sin(\vartheta-\pi/2)+ypy-2\,\mathrm{In}\left[(ny1+1)/2\right]\big)(-1)^{ny1},\\ zy &= \big((t-tpz)\sin\vartheta+zpy-2\,\mathrm{In}\left[(nzy+1)/2\right]\big)(-1)^{nzy}, \end{aligned}\right\} \quad (6.5a)$$

$$\left.\begin{aligned} z0 &= \big((t-tpz)\cos\vartheta+zpz-2\,\mathrm{In}\left[nz0/2\right]\big)(-1)^{nz0},\\ z1 &= \big((t-tpz)\cos\vartheta+zpz-2\,\mathrm{In}\left[(nz1+1)/2\right]\big)(-1)^{nz1},\\ yz &= \big((t-tpy)\cos(\vartheta-\pi/2)+ypz-2\,\mathrm{In}\left[(nyz+1)/2\right]\big)(-1)^{nyz}. \end{aligned}\right\} \quad (6.5b)$$

As seen from the equations, a probe point position on the wave front is defined now by the coordinates $\{y0, yz\}$, $\{zy, z1\}$ in the region of polar angle $\vartheta < \pi/2$, and by the coordinates $\{y1, yz\}$, $\{zy, z0\}$, when $\vartheta \geq \pi/2$.

Now we can apply the above results to investigate configurations of the perturbation wave front in a two-dimensional channel and in a pipe of square cross-section, assuming the test volumes of the plane channel being of the sizes $dx = dy = 8$, $dz = 1$, and of the sizes $dx = 6$ and $dy = dz = 1$ in the pipe. As to the two-dimensional channel, the wave front configuration is sought for the primary perturbation parameters $tp = 0$, $xp = 1.2$, $yp = 3$ and $zp = 1$ (the upper wall). In the pipe these data are supposed to be $tp = 0$,

$xp = 3$, $yp = 0.6$ and $zp = 0$ (the low wall). To be definite, we seek the configurations at the instant $t = 6.5$ (the flat channel) and at $t = 2.1$ (the pipe).

As has been shown in the beginning of the chapter, the wave front configuration for the main flow in the flat channel is computed by means of formulas (6.1) and (6.2). Program 6.1 can serve for computation and plotting of the wave configuration at the indicated conditions and $a = 0.03$.

**Program 6.1** Computing of a wave configuration in a flat channel at *pt*=6.5, *xp*=1.2; *yp*=3; *zp*=1

```
dx=8; dy=8; dz=1;
tp=6.5; xp=1.2; yp=3; zp=1;

n = IntegerPart[(t − tp) Cos[ϑ] + zp];
x = (t − tp) Sin[ϑ] Cos[φ] + xp; y = (t − tp) Sin[ϑ] Sin[φ] + yp;
z = ((t − tp) Cos[ϑ ]+ zp − 2 IntegerPart[(n+1)/2]) (−1)^n;
gr1= ParametricPlot3D[{x, y, z},{φ, 0, 2π},{ϑ, 0, π/2}, Boxed→True,
    PlotRange→{{0, 8},{0, 8},{0, 1}}, BoxRatios→{8, 8, 2},
    Ticks→{Automatic,Automatic,{0, 1}},
    AxesLabel→{"x", "y", "z"}];
```

The first line of the Program indicates the channel volume under investigation. The second line assigns quantities to parameters of the primary perturbation, to the origin instant and coordinates of the perturbation center. Formulas (6.1), (6.2) by which the wave front coordinates at the instant $t$ are computed are given in the next three lines. The last section of the program contains a set of parameters for formation of the image of the perturbation wave front for the chosen instant.

The same purpose in the pipe is attained by employing relations like (6.4) and (6.5). The pattern of computations is presented in Program 6.2.

**Program 6.2** Computing of wave configuration in a pipe of square section at *pt*=2.1, *xp*=3; *yp*=0.6; *zp*=0

```
dx = 6; dy = dz = 1;
tp = 2.1; xp = 3; yp = 0.6; zp = 0;

ny1=IntegerPart[t Sin[ϑ]Sin[φ] + yp];
ny0=IntegerPart[t Sin[ϑ] Sin[φ] + yp −1];
            nyz=IntegerPart[t Cos[ϑ];
nzy=IntegerPart[t Sin[ϑ] Sin [φ];
            nz1=IntegerPart[t Cos[ϑ] + yp];
            nz0=IntegerPart[t Cos[ϑ] + yp −1];
xy=(t Sin[ϑ] Cos[φ]) + xp;
y1=(t Sin[ϑ] Sin[φ] + yp − 2 IntegerPart[(ny1+1)/2] (−1)^ny1;
```

```
y0=(t Sin[ϑ] Sin[φ] + yp − 2 IntegerPart[[(ny0+1)/2] (−1)^ny0;
yz=(t cos[ϑ]+ zp − 2 IntegerPart[(nyz+1)/2] (−1)^nyz;
gr1=ParametricPlot3D[{xy, If [φ≤π, y1, y0], yz}, {φ, 0, 2π},{ϑ, 0, π},
  Compiled→True, PlotRange→{{0, 6}, {0, 1}, {0, 1})}, Boxed→True,
  Ticks→{Automatic, {0, 1}, {0, 1}}, AxesLabel→{"x", "y", "z"},
  AxesEdge→{{−1, −1}, {1, −1}, {−1, −1}}];
```

This program is prepared precisely in the same manner as Program 6.1, with only one difference. The wave reflections number is found by means of formulas (6.4), and the probe point coordinates are calculated according to equations (6.5). Outcomes of the computation can be seen in Fig. 6.2. The left picture shows the wave front configuration after six reflections from two walls of the flat channel. On the right is seen the front configuration after two reflections from four walls of the pipe.

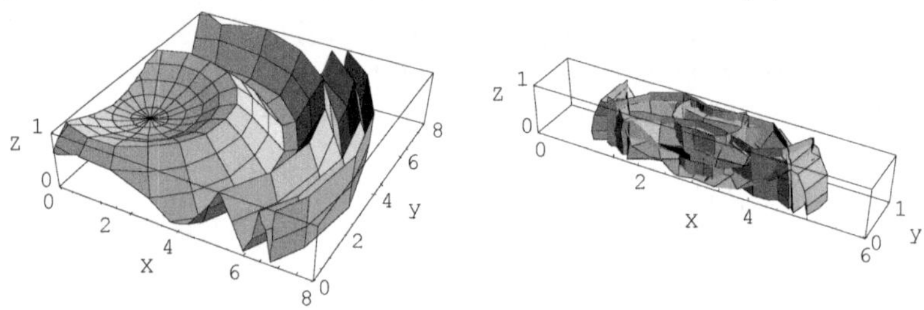


**Fig. 6.2.** A wave front configurations in a flat channel and in a pipe after six and two reflections respectively

As is seen from Fig. 6.2, the wave front in the flat channel after six reflections looks like a grooved surface with variable distance between the folds, and varying orientation of a normal to the wave front surface. Generally, orientation of the normal coincides with the direction of the velocity oscillations in the wave packet, and gives some rough notion about the structure of the oscillations in the flow. Hence one perturbation in the channel seems to create well-ordered oscillations of the flow parameters.

While propagating in the pipe, the perturbation wave front changes its spherical shape much faster. Herein a single wave creates in a short time such a perturbation configuration, that the simple source of its origin can hardly be guessed correctly. This appearance of perturbations is perceived not as an ordered structure, but rather as some random system of independent disturbances.

The pictures in Fig. 6.2 display the wave front in fixed instants of time. However a small improvement of Programs 6.1 and 6.2 allows us to describe the wave front evolution in time and simultaneous propagation of several waves. The pattern in Fig. 6.3 shows the picture of propagation of two waves arisen at different sites of a flat channel, and at different time instants. In fact, Fig. 6.3 demonstrates the process of superposition by the example of two waves.

Notice that the wave configuration in Fig. 6.3 resembles, to a certain extent, the image of one wave reflected from the walls, shown in Fig. 6.2. Looking at the pictures in Figs. 6.2 and 6.3, we may think that structures being formed by any sequence of the perturbation waves are of great interest.

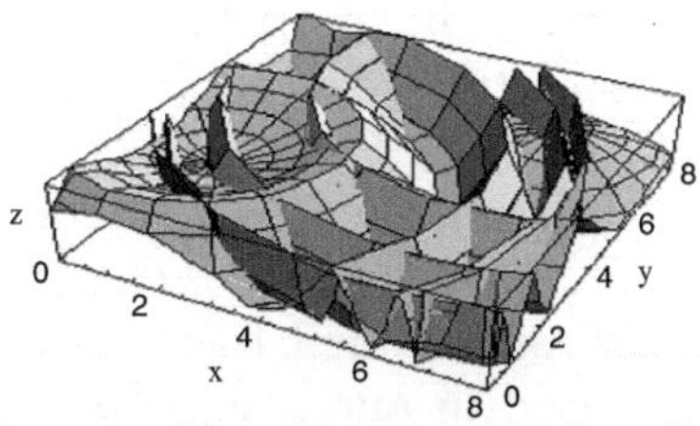


**Fig. 6.3.** Superposition of two waves in a flat channel

## 6.3 Structure of flow perturbations in channels

We proceed now to study the wave structures formed by a sequence of perturbations. All the waves originate on the surfaces of the walls, however they differ from each other by coordinates ***rp*** and instants *tp* of the origination. The structures are considered in the chosen conditions, i.e., in a flat channel and in a pipe of square section. The coordinates and instants of the primary perturbation origins are assumed to follow the law of random distribution. Some other laws will be discussed below in brief.

The process of structure formation is observed in some arbitrary though limited time interval T, including steady state of the structure, so the result does not lose generality. For this purpose, every wave is taken into account during all its lifetime $\tau$, while the wave amplitude exceeds, say, 1/40 of the initial amplitude (at the radius $a = 0.03$). Also, a certain mean rate K of the wave origin in dimensionless time units with the scale $d/c$ (or some quantity N of the perturbations arising during the time interval T) should be introduced to compute the structure. With these remarks we present a program for computing the perturbation structure in a flat channel.

**Program 6.3** Evolution of perturbation structure in a flat channel, $dx=dy=5$, $dz=1$

```
d=1; T=4; τ=2.8; SeedRandom[NUM];
xp=Table[Random[Real, (–1, 6)], {100}];
yp=Table[Random[Real, (–1, 6)], {100}];
zp=Table[Random[Integer], {100}];
tp=Table[τ– i(T+τ)/100, {i, 100}];
x=(t+tp) d Sin[ϑ] Cos[φ]+xp;
y=(t+tp) d Sin[ϑ] Sin[φ]+yp;
```

```
z=d(-1)^IntegerPart[(t+tp)d Cos [θ]+zp]×
  ×({(t+tp)dCos[ϑ]+zp-2IntegerPart[((t+tp)dCos [ϑ]+zp]+1)/2]);
Do[ParametricPlot3D[Evaluate[
  Table[If[0≤t+tp[[i]]≤τ,{x[[i]],y[[i]],z[[i]]},{0,0,0}],{i,1,100}],
  {φ,0,2π},{ϑ,0,π/2}, PlotRange→{{0,5},{0,5},{0,1}, Boxed→Truc,
  Ticks→{Automatic, Automatic, {0,0.5,1},{t,0,T,T/n}];
```

The program starts with the problem's initial conditions such as the channel height *d*, observation time T, wave lifetime τ, and law of the perturbation origins. The quantities of the places ***rp*** and instants *tp* of the perturbation sequences are generated as random numbers by command SeedRandom [NUM] of the system *Mathematica*, where NUM is usually a current time of day measured in small fractions of a second. However a fixed quantity has to be used in our case to provide the opportunity for structure evolution observation at the same probability distribution.

This basic number NUM = 29174083549372085208 is used in computation of distributions of the primary perturbation positions ***rp***(*xp*, *yp*, *zp*), and corresponding time instants *tp*, as the next four lines show. The expressions for the parameters include information concerning the channel volume where the perturbations arise, and the quantity of the perturbations N = 100, that provides the origin mean rate K = 25.

Coordinates *x*, *y*, *z* of any arbitrary point on the wave front surface for every initial position ***rp***(*tp*) are represented by formulas in the next lines. And in the end the command Do plots the images of all the wave fronts at one instant *t* after another within the range t∈[0,T]. As a matter of fact the program produces animation of the perturbation structure evolution with the step T/n.

The structure on the left in Fig. 6.4 demonstrates a frame from the picture series computed by Program 6.3. While copying the frame, the process of its duplication has been interrupted to show the structure from within. The picture on the right in Fig. 6.4 depicts the other example of the structure for rather different terms. Here are two sets of the determinative parameters used in computations for the left and right pictures:

$$dx=dy=5,\ dz=1\ \boldsymbol{rp}\in\{[-1, 6], [-1, 6], (0, 1)\},\ \mathrm{T}=4,\ \tau=2.8,\ \mathrm{NK}=25;$$
$$dx=dy=8,\ dz=1,\ \boldsymbol{rp}\in\{[-2, 10], [-2, 10], (0, 1)\},\ \mathrm{T}=10,\ \tau=10,\ \mathrm{K}=1\,.$$

Comparing the two pictures proves that the difference in the parameters becomes apparent due to the structure clear-cut distinction.

In the next chapter we shall return to the question. At present we consider evolution of the structure formed by a random sequence of perturbation waves in a pipe of square section. A pattern of the calculation can be constructed by using the technique employed in Program 6.2 for one wave propagation within the pipe, and in Program 6.3 for the sequence of waves in the flat channel. Combining these two programs results in Program 6.4.

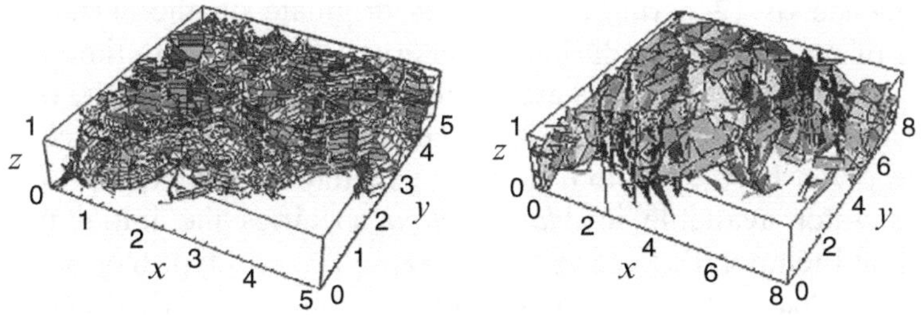


**Fig. 6.4.** Structures of perturbationsoo in a plane channel

**Program 6.4** Structure of perturbations in a pipe at wave sequence

```
d=3; dx=10; dy=dz=3; τ=1.2; N=12; t=0; SeedRandom[357204];
xyp=Table[Random[],{N/2}]; xzp=Table[Random[],{N/2}];
yyp=Table[Random[],{N/2}]; yzp=Table[Random[],{N/2}];
zyp=Table[Random[],{N/2}]; zzp=Table[Random[],{N/2}];
typ=Table[Random[],{N/2}]; tzp=Table[Random[],{N/2}];
ny1=IntegerPart[(t+τ typ) Sin[ϑ] Sin[φ]+yyp];
ny0=IntegerPart[(t+τ typ) Sin[ϑ]Sin[φ]+yyp−1];
    nyz=IntegerPart[(t+τ typ) Cos[ϑ]+yzp];
nzy=IntegerPart[(t+τ tzp) Sin[ϑ]Sin[φ]+zyp];
    nz1=IntegerPart[(t+τ tzp) Cos[ϑ]+zzp];
    ny0=IntegerPart[(t+τ tzp) Cos[ϑ]+zzp−1];
xy=d((t+τ typ) Sin[ϑ] Cos[φ])+dx xyp;
y1=d((t+τ typ) Sin[ϑ] Sin[φ]+yyp−2 IntegerPart[(ny1+1)/2])(−1)^ny1;
y0=d((t+τ typ) Sin[ϑ] Sin[φ]+yyp−2 IntegerPart[ny0/2])(−1)^ny0;
yz=d((t+τ typ)Cos[ϑ]+yzp−2 IntegerPart[(nz1+1)/2])(−1)^nyz;
xz=d((t+τ typ) Sin[ϑ] Cos[φ]+ dx xzp;
zy=d((t+τ typ) Sin[ϑ] Sin[φ]+zyp−2 IntegerPart[(nz1+1)/2])(−1)^nzy;
z1=d((t+τ typ) Cos[ϑ]+zzp−2 IntegerPart[(nz1+1)/2])(−1)^nz1;
z0=d((t+τ typ)Cos[ϑ]+zzp−2IntegerPart[ny0/2])(−1)^nz0;
grN=Show[ParanetricPlot3D[
    {xy[[i]], If[φ≤π, y1[[i]], y0[[i]], yz[[i]]},
    {xz[[i]], zy[[i]], If[ϑ<π/2, z1[[i]], z0[[i]]},
    {φ,0,2π},{ϑ,0,π }, Compiled→True, Boxed→True,
    PlotRange→{{0,10},{0,3},{0,3}, ViewPoint→{3.9,−7.2,3.6},
    AxesEdge→{{−1,−1},{−1,−1},{−1,−1}}, AxesLabel→{x,y,z},
    {t, 0, T, T/n}, {i,1,N}];
```

The program evaluates the perturbation wave structure by the time $t$ (in non-dimensional units scaled by $d/c$). To be definite, the structure is investigated in the pipe volume 3 × 3 × 10 cm at duration T = 1.2, when the sequence of 12 perturbation waves originate on the surfaces of two couples of opposite walls during the interval. The wave lifetime is taken to be equal $\tau = 1.2$. All the data are included in the first line of the program.

Origination of the waves within the perturbation sequence conforms to the rule given by the random number generator Table[Random[],{N/2}]. The generator available in *Mathematica* specifies the initial parameters (spatial and temporal), xyp, yyp, zyp, xzp, yzp, zzp, typ, tzp, separately in every try of "cast lots". The value of the generator seed NUM is maintained constant during the procedure for the same purpose and just in the same way as it has been done in Program 6.3. The program lines from the second to the fifth contain all the information.

It has been tacitly assumed that the primary perturbation has the characteristic dimension $a = 0.03$, certainly in dimensionless units. That means the boundary layer being thin. This rather small quantity predetermines the time interval T duration to get the structure steady state, and implies an opportunity to deal separately with the main part of the flow. So the quantities of the wave reflections for various combinations of the pipe walls, i.e., ny1, ny0, nyz, nzy, nz1, ny0, are computed by means of the next few lines of the program. Then the coordinates of the waves and reflected parts of the waves, xy, y1, y0, yz, xz, zy, z1, z0, are computed in the following eight lines.

The last operation grN assembles all the data by varying number i of the source waves in their development, and plots the configuration of the wave fronts by the desired instant $t$. The process of the structure evolution is illustrated by the frames shown in Figs. 6.5–6.7. All the pictures are plotted for the initial data indicated in Program 6.4.

Figure 6.5 represents the structure development process at its initiation. The figure makes it clear that the perturbation structure represents insulated wave fronts looking like quite a regular configuration. At the other extreme case shown in Fig. 6.6, the perturbation wave structure formed at a much later stage of the evolution, by the instant $t = \mathrm{T}$, results in a rather tangled picture. The wave configuration actually imitates a steady state of the perturbations, when the origin of new perturbation waves equalizes the deficiency caused by damping and dying of the propagating waves.

To obtain more complete notion concerning the inner structure, the view from within should be examined. To do that, a longitudinal thin stratum of thickness 0.2 cm extracted from the middle of the structure is shown in Fig. 6.7. The structure image reveals the wave distribution properties. The distribution scale is seen to vary in rather wide limits of the dimension. Note by the way that this range would be even wider if we had not omitted more weak waves from consideration.

The structure scale depends evidently on the mean quantity of waves simultaneously propagating in the chosen volume of the pipe, i.e., on the wave lifetime and specific rate of the wave generation. The structure dependence on generation rate is verified by the state of the stratum shown in Fig. 6.8. It is computed in the same conditions as in Fig. 6.7, though the perturbation specific density is three times as much this time.

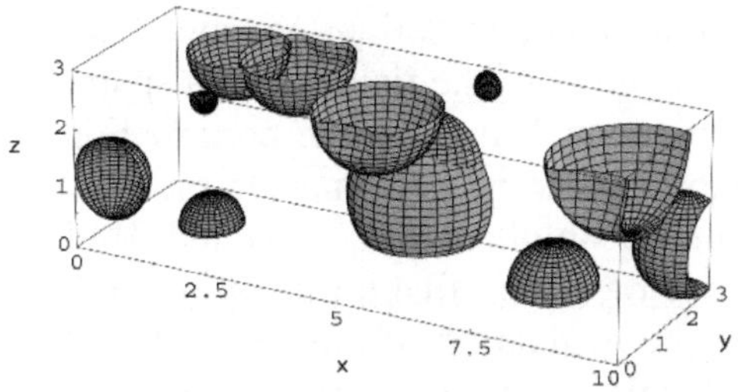


**Fig. 6.5.** Initial state of perturbation configuration, N=12

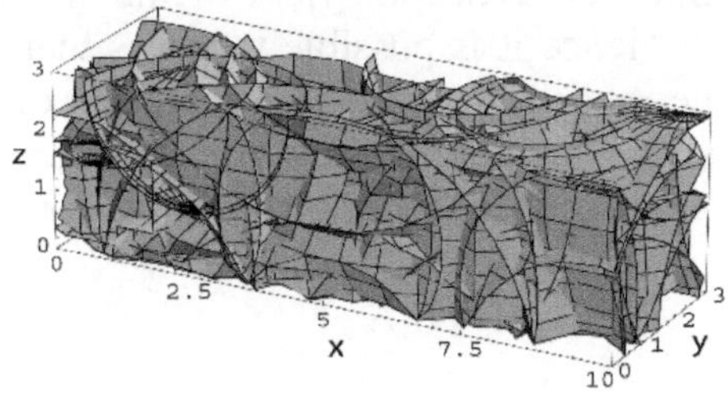


**Fig. 6.6.** Perturbation structure at instant t = T=1.2, N=12

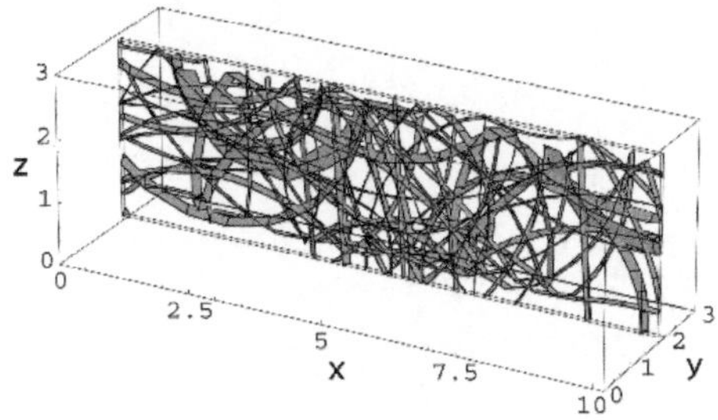


**Fig. 6.7.** Perturbation structure at instant $t$ = T=1.2, N=12, stratum 0.2 cm

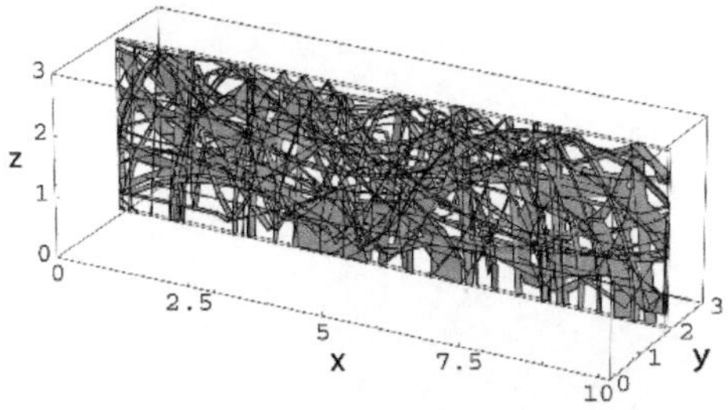


**Fig. 6.8.** The same structure, but for N=36

Anyhow, the images which we see on these frames represent patterns of allocation fields of the moving waves, registered at certain fixed instants. This field appears as a result of the linear propagation of acoustic waves. This field can have different structure depending on many factors, but it is calculable, when the initial perturbations are specified.

The perturbation waves of the field create a spatio-temporal field of the flow parameter oscillations. Naturally, the perturbation waves damp and die out in the end, being replaced by new ones. In the same way the oscillations gradually damp, but are simultaneously resumed. Both the parameter oscillations and dynamics of the oscillation damping and resumption have not been considered in the evaluation of the wave allocation field. Meanwhile the spatio-temporal field of the oscillations remains an important research problem.

At the same time, the allocation field of the waves gives an idea of their random nature. Hence it is possible to guess that the flow parameter oscillations might be random as well. The problems of basic value related to the study of the spatio-temporal field of the flow parameter perturbations and a process of the field chaotization will be considered in the next chapter.

# 7 Evolution of velocity oscillation field

## 7.1 Oscillations of flow parameters produced by a wave

As has been shown in Chapter 5, the primary perturbation propagates in the form of a wave packet which holds the oscillation of the hydrodynamic flow parameters. When the wave propagates, the oscillation amplitudes fade gradually. We consider the velocity perturbation $\boldsymbol{u}$ of the flow velocity $\boldsymbol{U}$ as an example. Generally the results refer to all the other parameters (5.6).

While a wave packet passes through some appointed position $\boldsymbol{r}(x,y,z)$ of a channel, the velocity oscillates at that position according to solution (5.2) of the wave equation (5.1). The velocity perturbation lasts at the test point $\boldsymbol{r}(x,y,z)$ during the time $2a/c$, as it follows from equation (5.2).

If a sequence of the wave packets propagates within the channel, the oscillations appear at the point $\boldsymbol{r}(x,y,z)$ in that succession as the packets pass through the point. It has been proved earlier that the perturbation amplitudes are small. Hence, as two wave packets or more continue to pass through this point, the perturbation total amplitude is defined by simple addition of $\boldsymbol{u}$-vectors, as the superposition principle says.

The resultant perturbation at that point means that the oscillation amplitude and orientation are time functions. Therefore the totality of such functions throughout the channel presents a vector field of parameter oscillations, which is a vector function of both the coordinates and time, $\boldsymbol{u}(x,y,z,t)$. In other wards, it is a spatio-temporal field of the amplitude vector of the velocity oscillating in every point of the flow under consideration.

This field will be studied below in detail. In the meantime we find the field which is created by a wave packet during its propagation through a space, to demonstrate the computing technique. Note, the vector $\boldsymbol{u}(x,y,z,t)$ is a parametric function, for its arguments depend on the path length $\ell$ that some element of the wave front area covers to arrive at the point $\boldsymbol{r}(x,y,z)$. In non-dimensional units it is the same quantity as the corresponding time interval $t$. The length in its turn is determined by formulas similar to equation (6.1).



L.N. Pyatnitsky, *Turbulence Nature and the Inverse Problem*, Fluid Mechanics and Its Applications 89, 85–102.

To find the three-dimensional field of function $\boldsymbol{u}$ we resort to the help of system *Mathematica* packages named <<Graphics`PlotField3D` and <<Graphics`ParametricPlot3D`. Then the following simple program computes the field.

**Program 7.1** Flow velocity oscillations produced by a wave

vel = PlotVectorField3D[{$\frac{x}{x^2+y^2+z^2}$, $\frac{y}{x^2+y^2+z^2}$, $\frac{z}{x^2+y^2+z^2}$},
{$x$, 0.5, 0.5}, {$y$, 0.5, 0.5}, {$z$, 0.5, 0.5},
PlotPoint→6, VectorHeads→True, Axes→True,
PlotRange→{{0,0.4},{0,0.4},{0,0.4}}, AxesLabe→{"x","y","z"},
AxesEdge→{{–1,–1},{1,–1},{1, 1}}, DisplayFunction→Identity];

sph = ParametricPlot3D[{0.1 Cos[$\vartheta$]Cos[φ], 0.1 Cos[$\vartheta$] Sin[φ], 0.1 Sin[$\vartheta$]},
{$\vartheta$, 0, π/2},{φ, 0, π/2}, DisplayFunction→Identity];

comb = Show[{vel, sph}, DisplayFunction→$DisplayFunction];

The program contains two special commands designed for use in problems like ours, that lighten the work of program text composing. One of them, vel, in the first indention, computes the vector field of the oscillation velocity amplitudes at six positions of the wave fronts. The vectors are depicted in the form of arrows located on the surface of the wave front at its consecutive displacement with step $\Delta\ell = 0.07$ ($\Delta t = 0.07$), beginning from $\ell = t = 0.1$.

The other special command, sph, in the second indention, describes the wave front itself at one of six positions, the perturbation initial diameter being $a = 0.03$. The wave front looks like a segment of a spherical surface. In the end, the amplitude field of velocity oscillations is combined with the computed wave front. The combined picture is shown in Fig. 7.1.

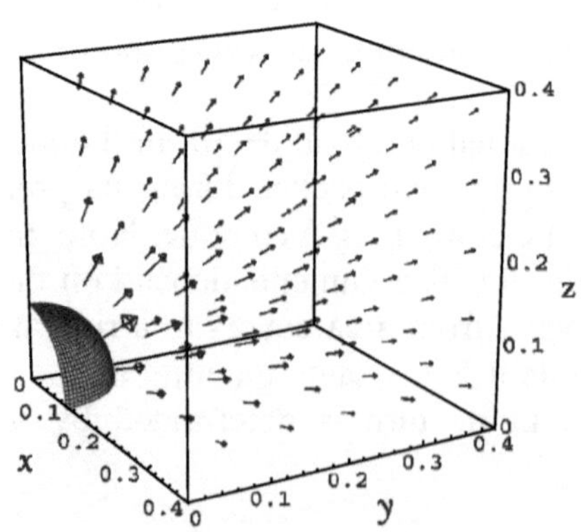


**Fig. 7.1.** Amplitude field of velocity oscillations at a wave propagation

The image space in the figure is limited by intervals of the direction angles $\vartheta \in [0, \pi/2]$, $\varphi \in [0, \pi/2]$ and the coordinate ranges $x, y, z \in [0.05, 0.4]$. The wave front is shown only for value $\ell = 0.1$ ($t = 0.1$). All the limitations have been introduced, lest any additional information in Fig. 7.1 would hide the main features of the field.

The spatio-temporal field of perturbations is interpreted on the basis of a wave model of the phenomenon. This field follows from the solution of, we emphasize, the linear wave equation (5.1) with a source function in the form of perturbations like those used in equation (5.2). The structure of the resultant field represents a joint configuration of the perturbation waves which are formed in the channel by means of relations of the type (5.13) and (5.14). When the main flow part is considered, these relations become much more simple, and are replaced by relations (6.1), (6.2) in the case of a flat channel, and by (6.4), (6.5) in a pipe of square section.

When we deal with a turbulent flow, the long-standing fundamental question arises, how the determinative exciting motion created by some given acoustic waves turns into random oscillations, which were once called fluctuations. The formation process of the irregular configurations of the wave fronts has been illustrated earlier in Figs. 5.2 and 6.5–6.8. Now when the oscillating velocity distribution can be computed, there appears an opportunity to understand what happens to the oscillations.

As we see from Figs. 6.6 and 7.1, evolution of the three-dimensional wave fronts along with amplitudes of the velocity oscillations may not be reproduced on a sheet of paper. Therefore the data on the configuration development in time (or as a function of path length $\ell$) may only be shown in some sectional plane of the three-dimensional process. Similar to Fig. 6.7 we consider the process in the plane $y = 0$.

The field in the plane is computed by a method similar to the one used in Program 7.1. Though, the evaluation of the sought functions in the plane is fulfilled in the system *Mathematica* by means of another additional software package, namely <<Graphics`PlotField`. As an example, Program 7.2 demonstrates a technique for computation of the velocity oscillation amplitudes in the flat channel plane $y = 0$ during one wave propagation.

We proceed from the assumption that the channel test volume is (in dimensionless units) restricted by the length $x \in [-3.5, 3.5]$ at the height 1. The primary perturbation looks like a hemisphere located on the lower wall, the radius being $a = 0.03$. The values and orientations of the oscillation amplitudes are computed for four distances covered by the wave front, $\ell = \{0.2, 0.9, 1.6, 2{,}3\}$ (the equality $\ell = t$ is valid in non-dimensional units). The picture area should be extended beyond the channel bounds to depict the velocity vector images in one scale.

**Program 7.2** Evolution of oscillation field produced by a wave

```
xp=0; zp=0; ℓ={0.2, 0.9, 1.6, 2.3}; N=IntegerPart[ℓ[[i]] Cos[ϑ];
lineU=Graphics[{Thickness[0.01], Line[{{-3.5, 1}, {3.5, 1}}]}];
lineD=Graphics[{Thickness[0.01], Line[{{-3.5, 0},{3.5, 0}}]}];
vec=Table[{{ℓ[[i]] Sin[ϑ] + xp, zp + (-1)^(N+zp) (ℓ[[i]] Cos[ϑ] -
    -2 IntegerPart [(N+1)/2])}, {(1-0.1)^N Sin[ϑ]/(2ℓ[[i]]),
    (1-0.1)^N (-1)^(N+zp) Cos[ϑ]/(2ℓ[[i]])}},
    {i, 4}, {ϑ, -(π/2)(1-0.1), (π/2)(1-0.1), 0.05π}];

gc1=Show[Graphics[{AbsoluteThickness[2],Circle[{0,0},0.3,{0,π}]},
     Frame→True, PlotRange→{{-5, 5},{-1.2, 2.2}},
     DisplayFunction→Identity];
gc2............................... ;
gc3............................... ;
gc4=Show[Graphics[{AbsoluteThickness[2],
    Circle[{0, -2}, 2.2, {0.36 π, 0.67π}],
    Circle[{0, 2}, 2.27, {1.147π , 1.33π}],
    Circle[{0, 2}, 2.245, {1.667π, 1.854π }],
    Circle[{0, 0}, 2.265, {0, 0.41}],Circle[{0, 0},2.24, {π-0.41,π}]},
    Frame→True, PlotRange→{{-5, 5},{-1.2, 2.2}},
    DisplayFunction→Identity];

 gr1=Show[[Graphics`PlotField`ListPlotVectorField[vec[[1]],
    HeadWidth→1, FrameTicks→{Automatic,{-1,0,1,2},{},{}},
    FrameLabel→{"x", "z", " ", " "}, PlotLabel→{"ℓ=0.2"},
    DisplayFunction→Identity], PlotRange→{{-3.5,3.5},{-0.2,2.8}},
    Frame→True, lineU, lineD, gc1, DisplayFunction→$DisplayFunction];
 gr2............................... ;
 gr3............................... ;
 gr4=Show[Graphics`PlotField`ListPlotVectorField[vec1[[4]],
    HeadWidth→1, FrameTicks→{Automatic,{-1, 0, 1, 2}, {}, {}},
    FrameLabel→{"x", "z", " ", " "}, PlotLabel→{"ℓ=2.3"},
    DisplayFunction→Identity], lineU, lineD, gc4, Frame→True,
    PlotRange→{{-5, 5},{-1.01, 2.01}}, Ticks→{Automatic,{-1,0,1,2}},
    DisplayFunction→$DisplayFunction];
```

When we deal with one wave, the coordinates *xp*, *zp* can be assigned at one's own choosing, and in the first line they are $xp = zp = 0$. Also the above mentioned instants are written in the line, and the number of the wave reflection is defined. Then two procedures lineU and lineD are introduced into the program to mark out the channel upper and lower borders, and the vector of the velocity oscillation amplitude, vec, is determined for four path lengths $\ell$.

Note, commands of type Do are not used in the program to show each separate step of the evaluation procedures. So the program's next lines compute configurations of the wave fronts, gc1, …, gc4 at the chosen instants, including reflected parts of the fronts. In the end, these data are employed in operations gr1, …, gr4 that construct the wave configurations. The field of the velocity amplitude vectors computed for the flat channel at the indicated four steps under the program is shown in the corresponding frames in Fig. 7.2.

The channel appears as a white stripe, and the space outside its limits is shaded. The wave fronts are represented by circle segments, and therefore radii correspond to the distances $\ell$ which are indicated in the right upper corner of each frame. Vectors of the velocity amplitudes are depicted by arrows whose length is proportional to the velocity amplitude.

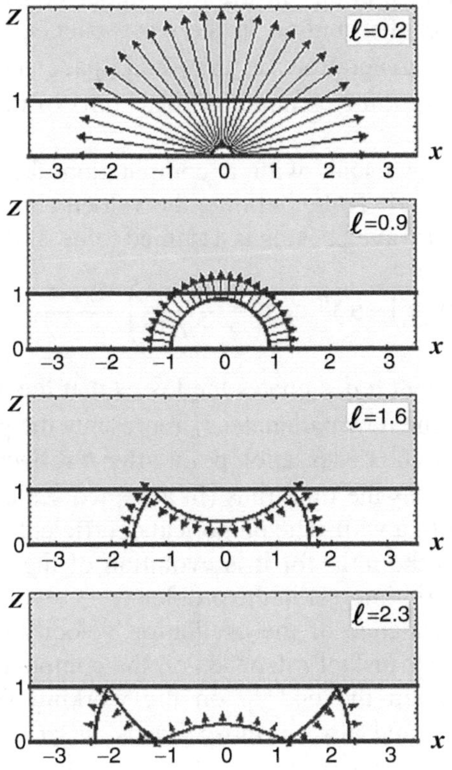


**Fig. 7.2.** Evolution of the velocity oscillation amplitudes, plane channel, $y = 0$

To study the field's special features we consider the quantitative characteristics of oscillations in the two-dimensional channel, analyzing properties of such dependences as the oscillation velocity amplitudes and spectra in time, u($t$), A[u($t$)], and in coordinates, u($\boldsymbol{r}$), A[u($\boldsymbol{r}$)]. The analysis of waves reflected from surfaces under these relatively simple conditions, such as in the flat channel, allows us to acquire important information about the field without applying cumbersome formulas peculiar to channels of complex geometry.

Let us begin a consistent consideration from the beginning, step by step. As we have seen, the parameter oscillations in time at some point $\boldsymbol{r}(x,y,z)$ of the flat channel are described by formulas (5.2) and (5.6) relating to the wave front surface (6.1), (6.2). According to (5.2) and (5.6), the oscillations exist, are not zero, only within the wave packet by the thickness $2a$, $|\ell_n - t| < a$. Therefore the oscillations appear at the point $\boldsymbol{r}(x,y,z)$ when the wave front arrives at the point, and ceases when the trailing edge of the wave packet leaves it. But the waves reflected from the walls pass through the point as does every original wave.

Generally, a description of the oscillation field needs knowledge of the data set on perturbing motions in the flow under consideration. The data include the parameter distribution in the wave packet, i.e., in the primary perturbations, $f_{r<a}(r)$, generation (in time and space) law for the primary perturbations, $\boldsymbol{rp}$ and $tp$, the channel profile, limits of the function $\boldsymbol{r}(x,y,z)$ variation, etc. [36].

We start first, as usual, with a computation for one wave packet moving in a flat channel. To be definite, the velocity amplitude (5.6) of the original and reflected wave packets is assumed to be of the form [36]

$$\mathrm{u} = (1-\alpha)^{\mathrm{n}} \cos\left(\frac{\pi}{2}\frac{\ell_n - \ell}{a}\right)\frac{\ell_n - \ell}{\ell}. \tag{7.1}$$

Here the coefficient $\alpha$ designates the losses that the wave incurs during each reflection, and the new parameter $\ell_n$ represents the path length that the wave covers to reach the observation point after $n$ reflections. The velocity amplitude diminishes as the time runs (in accepted scale $t = \ell$), and as the wave is reflected from a wall. The numerical coefficient in the expression is taken to be equal to the unit, for it is evolution of the velocity oscillation field, which is the main interest in the problem.

Now, time dependence of the oscillation velocity amplitude at some point $\boldsymbol{r}\ (x,y,z)$ and any instant $t$ depends on the computation time from the process beginning $t_1$ to the end $t_2$, on the working volume limited by parameters $\boldsymbol{r}_1$ and $\boldsymbol{r}_2$, and also on those places $\boldsymbol{rp}(xp,yp,zp)$ and times $tp$, where and when a perturbation arises. Hence the oscillation velocity amplitude at the observation point $\boldsymbol{r}\ (x,y,z)$ is a certain piecewise-smooth function depending on many variable quantities, which looks like

$$\boldsymbol{u}(\boldsymbol{r}_1,\boldsymbol{r}_2,t_1,t_2,\boldsymbol{rp},tp,a,\alpha,\tau,\boldsymbol{r},tp). \tag{7.2}$$

If all these data are known, the mentioned functions $\mathrm{u}(t)$ and $\mathrm{A}[\mathrm{u}(t)]$ can be computed for any point $\boldsymbol{r}\ (x,y,z)$ within a chosen part of the channel. Let the named parameters be as follows [44]:

$$\left.\begin{array}{l} a=0.1,\ \alpha=0.1,\ n=10,\ \boldsymbol{r}\{x, y, z\}=\{0.3,\ 0.5,\ 0.7\}, \\ \{tp, xp, yp, zp\}=\{0,\ 0,\ 0,\ 1\},\ \tau=10,\ \mathrm{N}=500. \end{array}\right\} \tag{7.3}$$

Here the quantity N relates to the computation time resolution which means that the wave lifetime $\tau$ is divided into N parts. According to Fourier transform rules, the quantity of the spectrum's resolvable elements is the number N halved. Evaluation of the sought functions u($t$) and A[u($t$)] is fulfilled by means of Program 7.3.

**Program 7.3** Oscillation velocity amplitudes and the spectrum, produced by a wave at conditions given by (7.3)

```
a=0.1; α=0.1; tp=0; xp=yp=0; zp=1; x=0.3; y=0.5; z=0.7;
ξ=x−xp; η=y−yp; ζ=2IntegerPart[(n+1)/2](−1)^zp+(z−zp)(−1)^n;
r=√(ξ²+η²+ζ²); ρ=r−(t−tp);
u=Sum[If[Abs[ρ]<a, Evaluate[(1−α)^n Cos[πρ/(2a)]ρ/r,0],{n,0,10}];
utab=Table[u,{t,0.2,10,0.02}];
gu=ListPlot[utab, PlotJoined→True, PlotRange→{−0.06, 0.06},
    Frame→True, PlotJoined→True,
    FrameTicks→{Automatic,{−0.6, −0.3, 0, 0.3, 0.6}, {},{}}
    FrameLabel→{"Time, N=20ν+1","Velocity, u(t) "}];
guf=ListPlot[Abs[Fourier[utab], PlotJoined→True,
    PlotRange→{{0, 250}, {0, 0.046}}, Frame→True,
    FrameTicks→{Automatic,{ 0, 0.015, 0.03, 0.046},{},{}},
    FrameLabel→{"Frequency, N=20ν+1", "Spectrum, A"}];
```

The first line of the program contains information on the initial data and location $\boldsymbol{r}_0(0.3, 0.5, 0.7)$ of that spot where the oscillations are examined. Then the wave position $r$ and velocity amplitude u, utab at various instants $t$, are defined. In the end the velocity gu and Fourier image guf of the velocity are computed and plotted within time interval $t\in[0.2, 10]$ with step 0.02. The computation output is presented in Fig. 7.3 by two graphs in which arguments of time $t$ and frequency ν are changed by the number N of the resolvable elements.

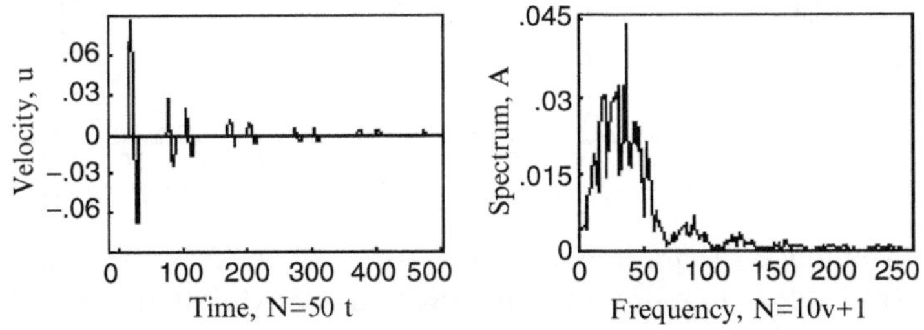


**Fig. 7.3.** Velocity oscillation amplitude u($t$) and Fourier transform A[u($t$)] according to data listed in (7.3)

By the example of one wave propagating in the channel, we can ascertain which parameters influence the oscillation spectrum in the channel and in what ways. It is evident that the form of Fourier spectrum does not depend on the wave start delay *tp* and location ***rp*** of the primary perturbation. However, the effect of the diameter *a*, the parameter distribution in initial perturbation $f(r < a)$, and of the loss factor α seems to be important.

Some preliminary information about the subject can be obtained from the function (7.1) represented in the form of a Fourier transform integral. Two terms of the function lead at integration to an output that contains an expression of sine integral function type, si. As is known, the function si tends to π/2 when the frequency *ν* belongs to the low-frequency part of the spectrum. On the other hand, the spectrum is approximated by the ratio cosψ/ψ in the high-frequency area. Note that the frequency ν should be replaced by product *a*ν in our case. Thus the spectrum under consideration has to be amplitude-modulated, as appearance of the spectrum shown in Fig. 7.3 proves, and also the frequency band has to broaden when the primary perturbation size *a* decreases.

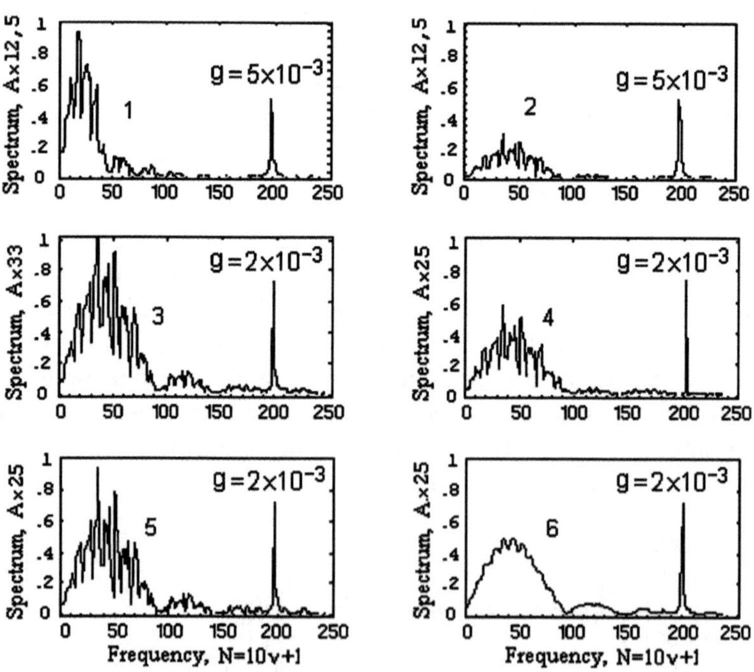


**Fig. 7.4.** Amplitude spectra in point ***r*** (0.3, 0.5, 0.7): $a$={0.2, 0.1} – spectra 1 and 2; $f=\{1-|r_n-t|/a\}$, $\{1-(|r_n-t|/a)^4\}$ – spectra 3 and 4; α={0.1, 0.5} – spectra 5 and 6

According to formula (7.1), amendments in the spectrum shape may depend on the function $f(r < a)$ and loss factor α which causes the wave damping and calls into existence the oscillation velocity jump at the wave reflection. At last, the spectrum frequency band changes under the influence of the test process duration T, while it has not come to the steady state. In this case the computation needs greater resolving power, i.e., the quantity N.

Influence of the parameters $a$, α values and the function $f(r < a)$ form on the spectrum can be estimated by varying them successively in Program 7.3. Some results of the variation are shown in Fig. 7.4.

When one of the parameters is varied, the others are imported from the data list (7.3). Spectra 1 and 2 are plotted for values of $a \rightarrow \{0.2, 0.1\}$; spectra 3 and 4 relate to the functions $f \rightarrow \{1\,|\,r_n - t\,|/a\}$, $\{1\ [(r_n - t)/a]^4\}$; and spectra 5 and 6 correspond to the values of coefficient $\alpha \rightarrow \{0.1, 0.5\}$. The separate spectral lines on the graphs are spectra of sine functions by amplitude g and frequency ν. These are added to the velocity for estimation of the absolute values of the spectral parameters depicted on the graphs.

As a comparison between curves 1 and 2 reveals, the frequency band of the oscillation spectrum is defined by the wave packet thickness $2a$, or the primary perturbation dimension. The whole spectrum contains a series of harmonics. The half-width of the harmonics each taken separately, and the distance between them, depend on the dimension $2a$ as well. The difference between the spectra 1–4 means that features of the harmonics and decrement of their damping are caused by a function of the $f(r < a)$ kind. Finally, it is clear from spectra 5 and 6 that increase of the energy loss factor at the wave reflection smoothes over the spectral curves.

## 7.2 Spatio-temporal field of oscillations in a wave sequence

When properties of the oscillations produced by one wave are known, we can turn to study of the oscillation field produced by many waves propagating in the channel. Certainly, the field is determined, if the start time *tp* and origin place ***rp*** of every primary perturbation in the sequence are given.

To compute the field, Program 7.3 should be extended to the case of the perturbation sequence including, say, eight ones. We intend to deal with the flat channel, the height and length being $z = 1$ and $x \in [-7, 7]$, when the observation region is limited by the length $x \in [-5, 5]$, accordingly. Let the instants and places of the perturbation origins obey the law of random numbers. The configuration of the oscillation amplitudes is computed for two instants, $t = 0.2$ and $t = 2.3$. The proper vector diagrams are demonstrated by two frames in Fig. 7.5.

Requirements for computation of the oscillation vector amplitudes shown by two frames in Fig. 7.5 correspond completely to the conditions in which the first and last frames in Fig. 7.2 were obtained, with the exception of the wave quantity. Comparison of the appropriate frames in the figures reveals a quick formation of the chaotic character of the velocity amplitude configuration of the frames in Fig. 7.5.

As is clear from the frames in Fig. 7.2, one wave in half-space creates a field of velocity oscillations which appear as an entirely definite and regular configuration. The wave reflections evidently change the direction and amplitude of the oscillation vectors in the channel and complicate the configuration. However the general effect of spatio-temporal field regularity does not vanish.

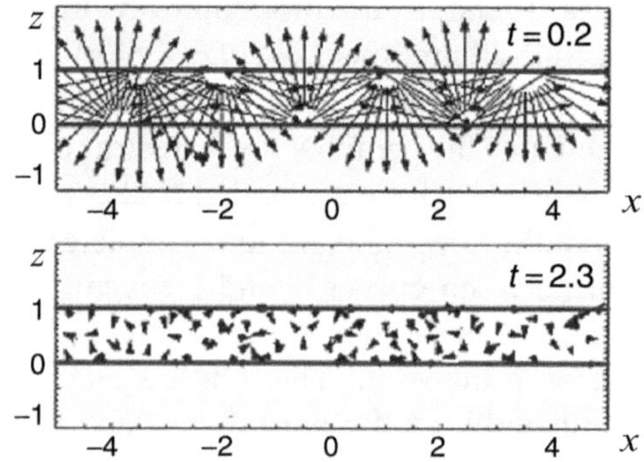


**Fig. 7.5.** Amplitude field of oscillating velocity in plane channel: propagation of eight perturbations, two time instants $t=0.2$ and $t=2.3$

As to the vector field of the oscillation amplitudes in Fig. 7.5, its configuration appears to be regular at the initial stage of the development (the frame $t = 0.2$), though the regularity is difficult to recognize at the later stages. When $t = 2.3$, only eight waves propagating in the channel (the wave fronts on the diagram are omitted) form the oscillation field that looks as if it is the result of chaotic motions [32, 35].

Note among other things, that despite the random origination of the perturbations forming the spatio-temporal vector field of the frame $t=2.3$, in Fig. 7.5, their dispersion as they arise takes a very small time interval, $tp \leq 0.5$. In fact, the initial perturbations appear almost simultaneously. The choice of such a small interval is stipulated only for the opportunities of a graphical data representation. For this reason the field chaotic state development concerns only a direction of the oscillation vectors on the graph.

In reality the interval $tp$ is naturally not limited. When the perturbations arise as some continuous sequence, the oscillation vectors differ not only by direction, but also by amplitude. So that quite deterministic process of propagation of the perturbations in a channel produces in a short time an image of random oscillations, which matches with the structure of wave fronts presented in Figs. 6.6–6.8.

Now we intend to investigate the function of the velocity $u(t)$ and the spectrum $A[u(t)]$, which can be observed in the point $\boldsymbol{r}(x,y,z)$ when many waves propagate in the channel. We assume, that each of the waves damps during its lifetime. In the development, the field structure passes through some stages in the evolution, until it reaches a steady state. Here and henceforward, steady structures are considered.

Let $k$ perturbations arise during the time $T$ when the flow in a steady state is observed. The perturbation wave is reflected $n$ times on average. Then the mean number of the perturbation waves simultaneously propagating in the channel will be equal to $kn\tau/T$, irrespective of the sequence form in which they appear. The value of the number means the process reciprocal resolution $\beta$ in time.

Definition of the structure stabilization time of the oscillation structure steady state is similar to the problem about the water level in a vessel at a constant average inflow and at an outflow velocity depending on altitude of this level. Similar to the case of the water level, the balance between origination and damping of the waves is defined by an expression of the type $\exp(-t/\tau)$.

According to the expression, the structure can be considered as satisfying a steady state condition at the instant $t_1 = 3\tau$, when the number of waves available in the channel does not differ from the dynamic equilibrium quantity more than 5%.

As to the experimental measurements, the parameter oscillations in a channel are used to record a time function at some chosen point, or an instantaneous distribution in some direction, often along the wall or in the transverse direction. We start with computation of the velocity oscillations in a fixed point.

To ensure the parameter oscillations being at dynamic equilibrium in the chosen point $\boldsymbol{r}(x,y,z)$, the computation time should belong to the following interval $t \in [t_1, t_2]$, where $t_1 = 3\tau$ and $t_2 \leq T$. For this purpose it is necessary to take into account all the perturbations originating at the distance from the spot of observation, which the wave covers during its lifetime $\tau$. For example, if the wave exists during $\tau = 10$, the test volume boundaries should lie in the limits

$$\{[x_1, x_2], [y_1, y_2], (z_1, z_2)\} \in \{[0, 20], [0, 20], (0, 1)\} \tag{7.4}$$

with centre in a plane $x = y = 10$.

At the same time, the survey of the channel volume surrounding the point under investigation has to respond to the requirement of the equilibrium state of oscillations as well. For this purpose it is necessary to consider all the perturbations which originate within the distance from the observation point, that a wave covers during its lifetime $\tau$.

Time properties of oscillations in the point $\boldsymbol{r}(x,y,z)$ can be computed by summing the velocity vector function

$$\boldsymbol{u}(a,\alpha,\boldsymbol{r}_1,\boldsymbol{r}_2,t_1,t_2,\boldsymbol{rp},tp,\boldsymbol{r}\{x,y,z\},\tau,T,k,\boldsymbol{r},t) \tag{7.5}$$

over all $k$ perturbations. We employ data of the example in computing the velocity oscillations in point $\boldsymbol{r}(10,10,z)$. To have the full set of initial conditions, let us assume that $a = \alpha = 0.1$, the time interval $[t_1, t_2] = [30, 50]$, coordinates of the observation point $\boldsymbol{r}\{x,y,z\} = \{10, 10, 0.7\}$, lifetime $\tau = 10$, the process is observed during $T = 50$, and the wave quantity $k = 100$. Under such circumstances the wave amplitude decreases for a lifetime approximately by 300 times.

Now the following program computes the velocity amplitude u($t$) and the velocity amplitude Fourier transform A[u($t$)] at the mentioned observation point.

**Program 7.4** Velocity oscillation amplitudes and the spectra in point $\boldsymbol{r}(10, 10, 0.7)$ for random sequences $\boldsymbol{rp}$ and $tp$

```
a=0.1; α=0.1; T=50; k=100; SeedRandom[NUM];
x=10; y=10; z=0.7; x1=0; x2=20; y1=0; y2=20;

xp=Table[Random[Real,{x1, x2}],{k}];
yp=Table[Random[Real,{y1, y2}],{k}];
zp=Table[Random[Integer],{k}];
tp=Table[T Random[],{k}];

ξ=x−xp; η=y−yp; ζ=2IntegerPart[(n+1)/2](−1)^zp+(z−zp)(−1)^n;
r=√(ξ²+η²+ζ²); ρ=r−(t−tp);

ut=Sum[If[Abs[ρ]<a, Evaluate[100(1−α)^n
    Cos[πρ[[i]]/(2a)]ρ[[i]]/r[[i]],0],{i,1,k},{n,0,10}];
uttab=Table[ut,{t,30,50,0.01}];

gut=ListPlot[uttab,PlotJoined→True,PlotRange→{{0,2080},{−3.2,3.2}},
    Frame→True, FrameLabel→{"Time, N=100v+1","Velocity, u(t)"}];

guf=ListPlot[Abs[Fourier[utab], PlotJoined→True,
    PlotRange→{{0, 310}, {0, 4.6}}, Frame→True,
    FrameLabel→{"Frequency, N=20v+1","Spectrum, A"}];
```

This program, seemingly, does not require any comments. Indeed, assignment of the initial conditions, evaluation manner of the wave configurations, and data on the velocity function u($t$) and spectrum A[u($t$)] representation have just been considered in connection with Program 7.3. As to the parameters $\boldsymbol{rp}$ and $pt$ describing random distribution of the primary perturbations in space and time, the matter has been discussed earlier in comments to Program 6.3.

A small difference concerns the function u($t$) (ut in the program) which is evaluated as a sum over all the perturbations marked by index i. Also, arising and damping of the perturbations are maintained at constant equilibrium level on average, while the oscillation parameters are computed. At last, the time duration $T$ is divided into N = 2,000 equal intervals.

## 7.3 Chaotization of a spatio-temporal field

Hitherto we used random distribution of the initial perturbations in the sequences ***rp*** and *pt*. Meanwhile, as pointed out at the beginning of section 6.3 and in comments to Program 7.4, any other form of the sequences ***rp*** and *pt* may be employed in this sort of program. Then the chosen kind of sequences ***rp*** and *pt* has to replace appropriate procedures in Program 7.4. For instance, if there is a necessity to assign, say, the perturbation constant spacing in the sequence *xp*, we should substitute the operation Random [Real,$\{x_1, x_2\}$] for Range[$x_1, x_2, dx$].

The function u($t$) in non-dimensional time units and function A[u($t$)] in non-dimensional frequency $\nu$ are plotted in Fig. 7.6. On the left column are graphs of functions u($t$), on the right column are the spectra A[u($t$)]. The data have been computed by means of programs similar to Program 7.4. Three combinations of various kinds of sequences ***rp***, *pt* were used in the computation [36].

The perturbations following the random sequences ***rp*** and *pt* were formed by the operation Random. The frames with the data occupy the upper row in the figure. The next combination of sequences includes perturbations arising at five fixed positions uniformly situated along the length $\{x_1, x_2\}$, while the time instants follow a random distribution. These data are represented in the second row. And the last combination shows the perturbations arising at five fixed positions at regular time intervals equal to $T/20$. These functions are in the third row.

The time and frequency are presented in the figure by number of the intervals N = 100 $t$ and N = 20$\nu$ + 1. The spectra frequency band on graphs is limited by significant magnitudes of the velocity oscillation amplitude.

Thus the functional dependences plotted in Fig. 7.6 relate to oscillations at a fixed point $\boldsymbol{r}(x,y,z)$ and differ only in the type of sequences in which the perturbations arise. Once again, the velocity oscillation amplitude u($t$) and time spectrum A[u($t$)] on the graphs *a* correspond to the sequences ***rp*** and *tp* with random distribution of the perturbations. The graphs *b* represent the version in which the distribution *tp* is a random sequence, whereas perturbations in the sequence ***rp*** arise at fixed points.

Unlike the two previous alternatives, the graph *c* depicts the velocity oscillation amplitude u($t$) and the time spectrum A[u($t$)] for well-ordered distribution of the perturbations within the sequence ***rp***, when the period 0.05T is the main characteristic of the sequence *tp*.

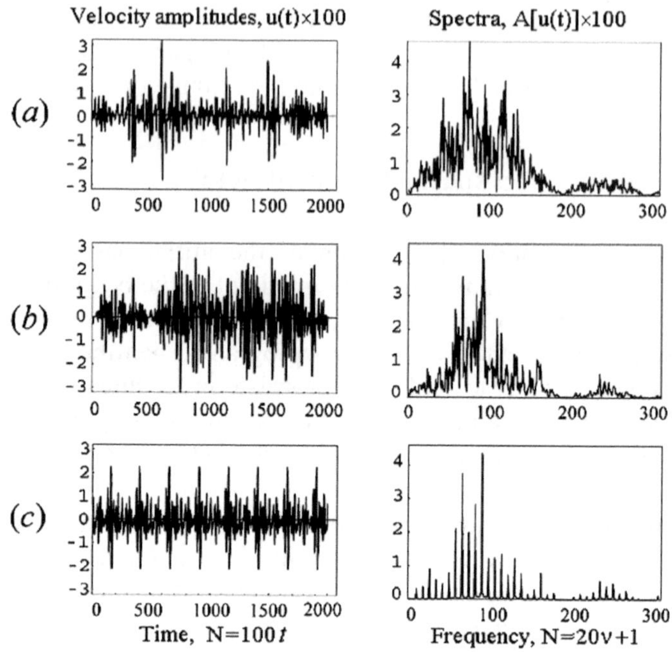


**Fig. 7.6.** Velocity oscillation amplitudes and the spectra at $x = y = 10$, $z = 0.7$ in time interval $t \in [30, 50]$: (*a*)−random ***rp*** and *tp*; (*b*)−5 fixed ***rp*** at *tp* random; (*c*)−5 fixed ***rp*** at periodical $tp = T/20$

Now we shall consider the structure of amplitudes of the velocity oscillations, i.e., space distribution of the oscillation amplitude and spectrum, u(***r***) and A[u(***r***)]. We assume that the structure develops in the same conditions as in the case of the time dependent functions u($t$), A[u($t$)]. Let the test segment of the straight line parallel to the $x$-axis be of length $L$ and pass through the point $\{x,y,z\} = \{10,10,0.7\}$, so that $L \in [0, 20]$. The process of the structure formation is considered at the time instant $t = T = 50$.

As before with the time dependences u($t$) and A[u($t$)], the spatial features of the oscillations on the straight line segment $L \in [0, 20]$ can be computed by summing up function ***u*** over all $k$ waves at every point of the line:

$$\boldsymbol{u}(a, \alpha, \boldsymbol{r}_1, \boldsymbol{r}_2, t_1, t_2, \boldsymbol{rp}, tp, \tau, T, k, L, \boldsymbol{r}) \tag{7.6}$$

The segment $L$ is located in the same volume, as in the previous computation. However, it is necessary to take into account the end effect in the case of spatial distribution. This means that the volume where the perturbations originate should widen at each end of the segment by a distance equivalent to the wave lifetime $\tau$, and the quantity of perturbations $k$ should be enlarged in the same proportion. Hence for the chosen conditions when $\tau = L = 10$ the new effective volume is to have the following limits [36]:

$$\{[x_1, x_2], [y_1, y_2], [z_1, z_2]\} \in \{[0,40], [0,40], [0,40], [0,1]\}. \quad (7.7)$$

The program for computing the sought dependences of the velocity oscillation amplitude u($x$) and the spectrum A[u($x$)] on the segment length $x$ is similar to Program 7.4. In this way we compute these two functions, u($x$) and A[u($x$)], within the length $x \in [10, 30]$ with the step $\delta x = 0.01$ at the time instant $T = 50$. These functions are shown in Fig. 7.7.

By analogy with Fig. 7.6, the left graphs show the velocity amplitude u($x$), and the right graphs relate to the spectra A[u($x$)]. The length in the direction of the segment $L$ and the spatial frequency in the direction are expressed by the number N, accordingly N = 100$t$ and N = 20ν + 1. The chosen resolution β meets the spatial spectral range N = 1,000 and the frequency band ν = 50. However the spectra on the graphs are limited by the range of ν ≈ 15, since the spectral density beyond the range is found to be too small.

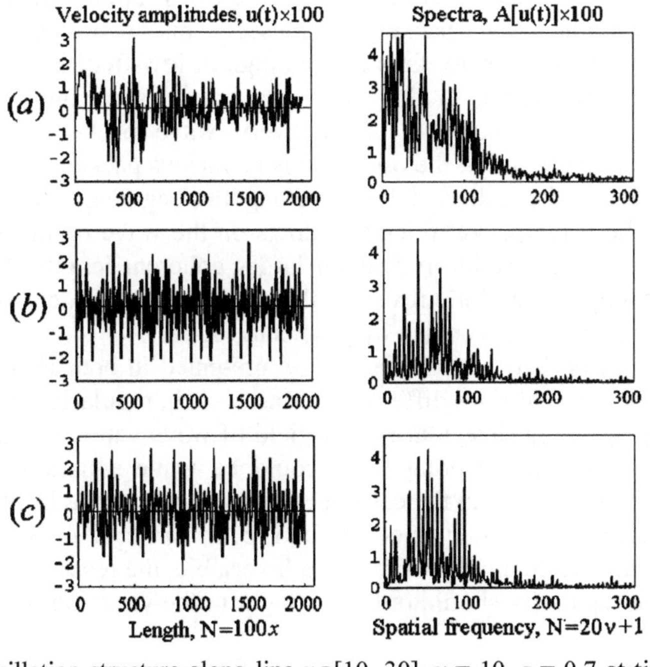


**Fig. 7.7.** Oscillation structure along line $x \in [10, 30]$, $y = 10$, $z = 0.7$ at time instant $T = 50$: ($a$)−***rp*** and *tp* random; ($b$)−5 fixed ***rp*** at *tp* random; ($c$)−5 fixed ***rp*** at periodical $tp = T/20$

Graphs $a$, $b$, $c$ in Fig. 7.7 correspond to just the same sequences ***rp*** and *tp* as in Fig. 7.6. Comparison of the graphs in Figs. 7.6 and 7.7 shows that temporal and spatial functions of the oscillations have similar characteristics. For instance, the curve that envelops the spectrum has a maximum which corresponds to the frequency ν ≈ $1/2a$ = 5 in both cases. Also there can be seen separate lines in the spectra, and the first more or less strong line is at the frequency $\nu = 2$ which coincides with the mean frequency of the perturbation that arises.

At the same time, there are two essential distinctions between the spectra presented in Figs. 7.6 and 7.7, concerning the spectral density distribution and contrast of the spectral lines. Density of the spatial spectra in low-frequency (long-wave) range exceeds considerably density of the temporal spectra. This distinction is a quite explicable effect.

Actually, the reference frequency of oscillations depends on the thickness $2a$ of the perturbation wave packet. The spectrum density at frequencies other than that depends on the coherent length of the oscillations in the wave trains taking part in superposition. At a fixed point, when the temporal spectrum is investigated, the coherent length may exceed the quantity $2a$ due only to synchronization of the oscillations in the wave packets passing through the observation point one after another.

The probability of such an event diminishes rapidly when the density of the available wave packets decreases. This is the reason why the temporal spectrum density falls in the range of low frequencies, as Fig. 7.6 shows. The coherence in the case of a spatial spectrum is a result of intersection of the segment $L$ by every wave packet, in addition to synchronization between the oscillations in various packets.

Therefore, the coherent length along the segment $L$ depends not so much on the wave packet thickness $2a$ as on the wave radius $R$ and on the crossing angle. It results in a much larger coherent length of oscillations along the length $L$, and in growth of the spatial spectrum density, especially at low frequencies, as verified by graphs in Fig. 7.7.

Comparison between graphs *a, b, c* presented in both Figs. 7.6 and 7.7 reveals, that spectral lines differ in contrast, i.e., in regularity of oscillations. Generally speaking, irregularity of a field of wave nature can be stipulated for various reasons. Among them could be wave equation nonlinearity, channel shape (e.g., as in the case of Sinai billiards), wave front profile, and also properties of the sequences ***rp*** and *tp*. In our case of the linear wave equation and the channel profile being invariable, the reason for a change in oscillation regularity should only be sought in the variation of sequences ***rp*** and *tp*.

As we can see from graphs *a* in Figs. 7.6 and 7.7, where the sequences ***rp*** and *tp* are random, the oscillations in both cases have continuous spectrum, though some separate spectral lines are visible against the continuum background. The graphs *b* in the figures, where sequences ***rp*** are ordered whereas sequences *tp* remain random, show that frequency ranges of the spectra shorten, and the separate spectral lines are given a better contrast.

The graphs *c* obtained at the full ordered sequences ***rp*** and *tp* discover the separate lines dominating in the spectra. The spectrum *c* in Fig. 7.6

consists of 23 separate lines, and the mean frequency shift of each of the next lines amounts to about $\Delta\nu = 0.4$. Note, the shift corresponds to a mean arising rate of the perturbations. In fact, this spectrum represents velocity periodical oscillations at the point under observation.

Thus, the degree of order in the oscillations depends on the structure of the perturbation sequences in space and time, ***rp*** and *tp*. There exist two extreme alternative sorts of sequences. Random sequences ***rp*** and *tp* entail in every respect a disordered spatio-temporal field of velocity oscillations, and evidently of other parameters of the flow.

On the other hand, the well-ordered sequences ***rp*** and *tp* involve a regular spatio-temporal field of the parameter oscillations, so that the field at some observation point proves the mode of usual acoustic oscillation. Note by the way, that the oscillations at neighboring test points of the field are different. Therefore, the field gives an impression of chaotic oscillations, even if the sequences ***rp*** and *tp* are entirely ordered, when it is observed with the unaided eye.

Now, a chaotic oscillation field may be found in either the random or the stochastic state. These qualities of the field are not discernible superficially, for the oscillations in both cases have continuous Fourier spectrum and a diminishing autocorrelation function. In essence, however, the random process differs from the stochastic one in the respect that it is generated by a system with an unlimited number of degrees of freedom, or depends on a random factor, e.g., on random initial conditions.

This delicate problem is worth special emphasis. As a matter of fact, we have seen earlier that the arbitrary initial phase $\beta$ in the formulas (1.15) and (1.21) for a small perturbation looks like a random factor which is similar to the random sequence *tp* applied to the time origination of the perturbations in our conception. Meanwhile, the sequences ***rp*** are the same as the sequence *tp*, in the sense that it relates to origination of the random perturbations. In other words, the random factor $\beta$ is just the same as the sequences ***rp*** and *tp*, as made obvious in the foregoing.

It has been found that important results concerning many observations made with turbulent oscillations can be explained on the basis of the wave conception, starting with Reynolds' experiments with a colored streamlet, including the turbulence intermittency phenomenon, and finishing with the Kolmogorov–Obouhov law concerning local properties of turbulence. Here we will touch upon the wave approach to that law.

The structure of flow velocity oscillations, with scale $l$ being small in comparison with the flow diameter, is locally homogeneous and isotropic. Figs. 7.5 and 7.7 confirm this supposition. The latter property means that the oscillating motions, within region with dimensions of the order $l$, do not depend on the direction of the mean flow velocity, and the oscillation features are completely independent of the direction. The problem in question implies the relative motion of the fluid particles in that region, but not the absolute motion of the region as a whole, i.e., the oscillation structure concerns the image and likeness of the turbulent flow.

One of the parameters that characterizes turbulent flow is the energy $\varepsilon$ dissipated per unit time and per the fluid mass unit. The quantity $\varepsilon$ is the energy flux by nature, which passes from the structural cells of larger scale to the ones of smaller scale down to the molecular level where the dissipation occurs due to viscosity. The structure of the flow oscillation field, as we have seen in §6.3 and in this chapter, is formed by superposition of the wave packets simultaneously propagating in the channel. The specific energy of a traveling acoustic wave is proportional to the local oscillation velocity squared, $E \sim \mathrm{u}^2$. Then the specific energy of the oscillations in the cell by scale $l$ can be expressed in the form $E_l \sim \mathrm{u}_l^2$ [3, §65].

In its turn, the energy dissipation is determined to be the product of the specific energy gradient and the effective viscosity, so that

$$\varepsilon = (dE_l/dl)\nu_{\mathrm{eff}}. \tag{7.8}$$

The gradient for small distances may be written as $dE_l/dl \approx (\mathrm{u}_l/l)^2$, and viscosity is estimated as $\nu_{\mathrm{eff}} \sim l\mathrm{u}_l$ [3, §33]. Substitution of these quantities into formula (7.8) gives the relation

$$\varepsilon \sim \mathrm{u}_l^3/l, \tag{7.9}$$

whence follows

$$\mathrm{u}_l = (\varepsilon l)^{1/3} \tag{7.10}$$

This is the well-known Kolmogorov–Obouhov law.

The analysis of hydrodynamic equations carried out above has shown that fluctuations of a turbulent flow can be interpreted as the oscillations of acoustic nature, which are formed by acoustic wave packets propagating in a channel. The spatio-temporal field of oscillations is a result of the wave packet superposition. According to the expounded representations, the model of the turbulent oscillations is a consistent consequence of well-known equations. Nevertheless, as a fundamentally new approach to the phenomenon interpretation is employed, the turbulence model that is considered needs experimental confirmation. This is the purpose of the next chapter.

# 8
# Experimental substantiation of turbulence wave model

Some fundamental assertions underlying the wave model of turbulence were summarized in Chapter 1. None of them, taken alone, seems to be unusual. However new effects of their combination may raise some doubts. Those assertions that need experimental verification first of all are: the simple wave structure in a channel and the character of compression waves transmitting momentum to the fluid, origination of acoustic perturbations, and formation mechanism of secondary waves of spherical type.

Certainly, a degree of conformity between the oscillation spatio-temporal fields computed by the offered method and observed in experiments should be verified as well. Here we will consider the experimental data relating to these problems.

## 8.1 Structure of a simple wave

The structure of a simple wave and accompanying pressure oscillations were observed in a pipe when a gas moved in front of the flame. It is common knowledge (see, for example [48–50]), that thermal expansion of combustion products produced by the flame propagating from the closed end of the pipe acts as a driving piston.

The flame operation analogy to the piston effect is explained in the following way (see, e.g. [49]). The combustion induces increase in temperature and pressure within the flame. The pressure excess propagates as compression waves moving forward in undisturbed gas, and backward in the burnt gas. The latter has high temperature, and the wave moves in it at large velocity. The waves reflected from the closed end of the pipe quickly equalize the pressure behind the flame. Therefore the burnt gas remains practically motionless, whereas the pressure waves moving forward produce the flow in front of the flame in the same way as the piston does.

The pipe of square cross-section with a quadrate side being equal to $d$ = 2.9 cm consisted of some interchangeable sections with a polished interior surface, including nests for pressure gauges. Some details will be sited below. One of the sections, the visual section, had two through windows of 19 cm length each, covered with plates manufactured from



L.N. Pyatnitsky, *Turbulence Nature and the Inverse Problem*, Fluid Mechanics and Its Applications 89, 103–124.

optical glass. Optical perturbations in the gas were observed through the windows by the Schlieren method.

In the experiments were recorded: the gas pressure in time at different points of the pipe, the process development by slit scan and high speed shooting of the Schlieren-image of the gas density perturbations. Mixture $CH_4 + 4O_2$ was ignited at a closed end of the pipe by a weak inductive discharge initiated by a spark in a gap between two thin electrodes spaced 4 mm apart. The technique of measurements of different parameters, details of the procedure and requirements are described in [51–53].

Examples of the pressure oscillograms registered in the pipe sections removed from the spark for distances $x = 0, 18, 35$ and $53$ cm are given in Fig. 8.1 [35]. Heavy line has been drawn as function $x(t)$ where $x$ is determined by the pressure sensor positions in the pipe, and time $t$ corresponds to the maxima of the sensor readings. Thus the line reproduces the flame front trajectory.

At the initial stage of the flame propagation, within the pipe's approximate length of 18 cm, the flame is seen to accelerate. At the acceleration stage the flame continues to fill up the pipe visual section, and the volume of the hot gas widens from the burning zone in all directions during that time. On the next part of the path, up to 53 cm, the hot gas begins to expand in one direction, along the pipe axis. The flame moves from that instant almost at a constant velocity, while the combustion acts as a piston pushing forward the gas in front of the flame.

The process development in this stage, where $x \in [18, 37]$, is presented in Fig. 8.2 by the slit scan of the Schlieren-image. Details of the process development are shown by frames of the rapid Schlieren-filming whose fragments are reproduced in Fig. 8.3. To magnify the spatial resolution of the process, the field of vision in the frames has been restricted by half of the visual section window, which length amounts to about 9 cm.

The distances from the ignition place to a current point, the pipe section $x$, are indicated on the photos; time from the instant of the explosive mixture ignition is set on the right of the frames. Though the photos show a projection of the structure in the vertical plane, because of a Schlieren technique peculiarity, the frames allow us to estimate the main features of the process. Now let us consider structures of the flame and of the gas moving in front of the flame at different stages of the process evolution.

Frames *a* in Fig. 8.3 are exposed at 58 and 146 μs after the ignition. These photos disclose development of the primary perturbation produced by the weak inductive discharge. The perturbation moves in the pipe at the sound velocity $c = 345$ m/s. It has the form of a spherical wave packet by thickness 4 mm, just equal to the initial discharge diameter. The spherical packet is reflected inevitably from the pipe walls, and superposition of the reflected waves forms some tangled structure of the gas density perturbations.

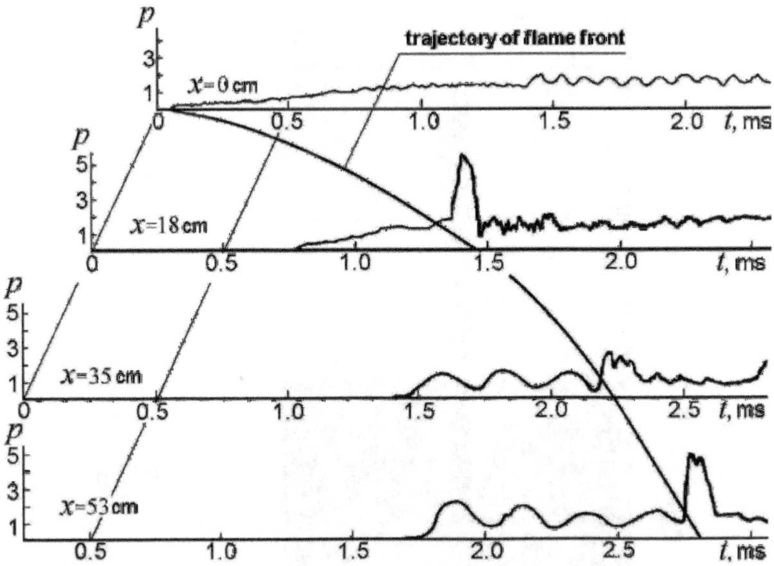


**Fig. 8.1.** Pressure oscillograms in pipe sections at $x = 0$, 18, 35, 53 cm

**Fig. 8.2.** Slit scan of process development in a pipe of square cross-section within the length 18–37 cm

The flame rapidly takes its contour of elongated shape, as frames *b* in Fig. 8.3 show. Photos *b* are exposed at 643 and 790 μs after the ignition. The wave of initial perturbation covers a distance of about 25 cm by the time instant, and disappears soon from the field of vision.

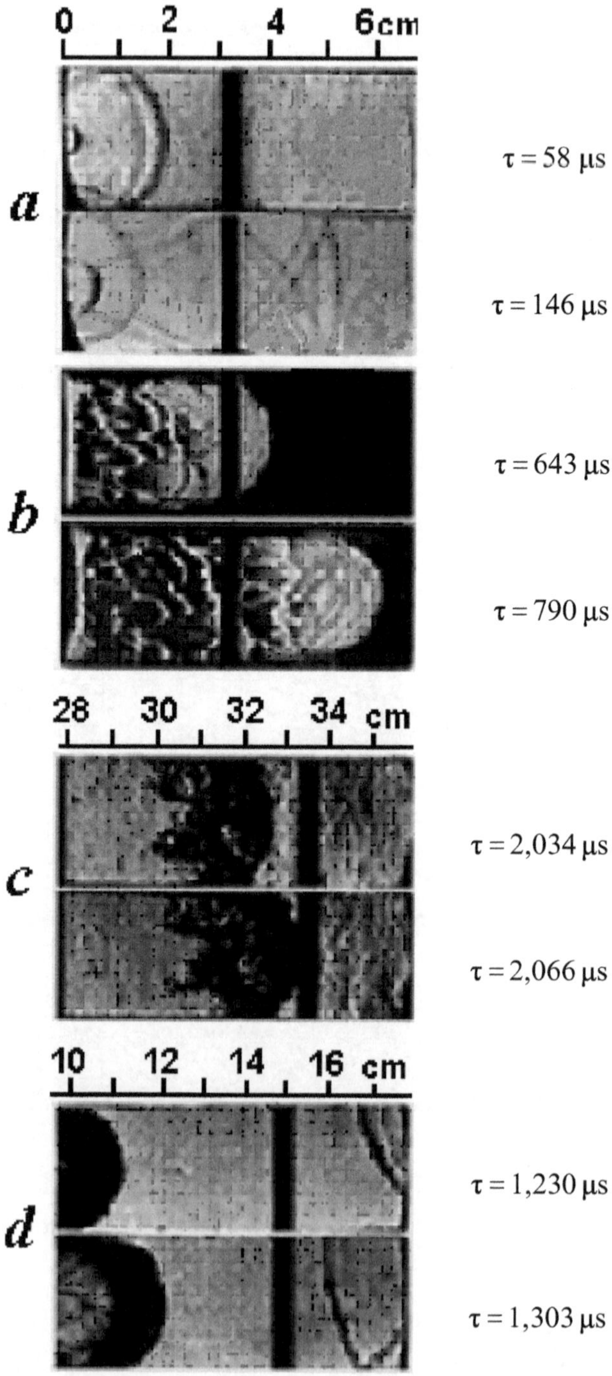


**Fig. 8.3.** Flow pattern and flame; τ – time delay of exposure after ignition

The frames *b* make it clear that the flame becomes of clearly expressed cellular structure. Cellular flames have been known for a long time, and have been observed in different experimental conditions (see elsewhere [54]). However the reason for appearance of this kind of structure remained unknown for some time. Eventually it was ascertained that these structures are formed under the action of waves of acoustic type [47, 55]. The frames *a*, *b* in Fig. 8.3, as well as others that are not shown here, definitively confirm this cause of cellular structure formation.

In the case under consideration, the amplitude of the wave covered the distance of 25 cm becomes 60 times less, and the wave density gradient oversteps the limits of the system visualization sensitivity. However the flame surface reacts to the gas's slightest motions. Therefore, the flow perturbation structure formed as a result of superposition of the acoustic waves is reproduced in the flame surface structure.

Generally speaking, the influence of acoustic waves on flame structure becomes apparent when the oscillating velocity of a medium in the wave amounts to a small share of the flame normal velocity. Since the normal velocity is known in our case, it is equal to 2.3 m/s for a given experiment, the flame structure can serve as a tool for quantitative estimation of disturbances in the flow running before the flame. In this way we can conclude that oscillations of the gas velocity in front of the flame are of the order $u/c \sim 10^{-4}$. The same can be said concerning the pressure oscillations, $p/P = u/c \sim 10^{-4}$.

Further development of the structure is demonstrated by frames *c* in Fig. 8.3, where are represented two quasi-stationary states of the already formed system 'piston–simple wave'. The inspection area in these frames extends over the pipe length $x \in [18, 37]$ cm. The flame appears in the area approximately in 1.8 ms after ignition, when the length of the simple wave amounts to 37 cm. The flow just in front of the flame is characterized in the pipe by Reynolds number $\mathrm{Re} \approx 2 \cdot 10^4$.

To gain more understanding of the structure within this stage, we shall compare patterns of flow represented in the forms of the slit scan (Fig. 8.2), instantaneous photos (frames *c* in the Fig. 8.3), and oscillogram in the pipe at $x = 35$ cm (Fig. 8.1). As is seen from Fig. 8.2, while propagating along the length $x \in [18, 37]$ cm, the flame does not change the shape, and moves at practically constant velocity. However, there are a lot of compression waves propagating in front of the flame.

The frames *c* in Fig. 8.3 show that superposition of the waves does cause the complex structure of perturbations in the flow preceding the flame-piston, and in the flame itself as well. It means that some discrete perturbations of finite intensity arise in the flow preceding the flame from time to time.

Structures available on the frames *c* in Fig. 8.3 allow us to make an important supposition concerning localization of the perturbation sources.

The perturbations seem to arise in the flow in front of the flame within a narrow ring which tightly sides with the pipe walls. To test this supposition we turn to the process slit scan shown in Fig. 8.2. Each perturbation wave moving before the flame leaves on the Schlieren photo a track in the form of a slant line whose declination is usually accepted as the propagation velocity.

However on closer examination, the lines are found to consist of separate segments by length from 8 to 16 cm each (2.5–5.5 calibers). Moreover, the image sharpness at the ends of the wave track segments becomes blurred and bends. Measurement of the angles at the segment ends gives the seeming velocity of the wave being equal to ~800 m/s at the beginning of the segment and equal to ~400 m/s at the end. Subtracting from these values the flow velocity which in the section $x = 35$ cm makes up ~50 m/s, we obtain accordingly 750 and 350 m/s.

The greater of these velocities might be observed only in the case of a shock wave at pressure jump $P_2/P_1 = 5.8$. However, according to the pressure oscillogram shown in Fig. 8.1, at the pipe section $x = 35$ cm, the pressure transducer reading does not exceed the level of the order $P_2/P_1 = 1.5$. This pressure jump level corresponds to the wave velocity which by no means is more than about 400 m/s.

This paradoxical result, at first sight, may serve as a definite proof that the perturbation waves arise at surfaces of the pipe walls. We shall consider a model of the pipe of square section with the side length $d = 3$ cm. The wave projection on the plane $y$, passing through the perturbation wave center, provides the best conditions for the wave front visualization. The wave starts on the lower wall at $x = z = 0$ just before the flame-piston model. The wave propagates in all directions, but we are interested in the area where $x > 0$, for the Schlieren method does not register small perturbations in the area $x < 0$ because of high temperature of the burnt gas mixture.

Let us determine a trace of side view of the wave front, which is seen through the pipe slit image of 19 cm long, just as in the experiment, during the period of $t = 6.7$ (the time is measured in the units of d/c). The wave is reflected from the horizontal walls six times during the period. The wave propagation trace on the slit scan is computed by means of Program 8.1.

**Program 8.1** Wave trace in plane $y = 0$ through spit for $px = 0$, $pz = 1.5$cm, $t = 6.7$

```
d=3; dx=0; n=6.7;
g6=ParametricPlot[{tdSin[ϑ]+dx,
    (tCos[ϑ]−2IntegerPart[(IntegerPart[tCos[ϑ]+1)/2])×
    ×d(−1)^IntegerPart[tCos[ϑ]]}, {ϑ,0,π/2}, Frame→True,
    PlotRange→{{0,20},{0,3}}, FrameLabel→{"x,cm","z,cm"},
    DisplayFunction→Identity];
```

```
line=Line[{{0.5,1.4}, {0.5,1.6}, {19.5,1.4}, {19.5,1.6}}];
combi=Show[{g6,Graphics[line]}, AspectRatio→1/5.7,
    DisplayFunction→$DisplayFunction];
```

In this simple program the wave trace in the slit field of vision is computed by the operation g6. Reflections from the vertical walls are ignored for the reason already mentioned above. The pipe and the slit contours are defined by command 'line'. The operation 'combi' prepares data and illustrates the wave trace in the chosen volume. The outcome of the calculation is reproduced by diagram in Fig. 8.4 [44].

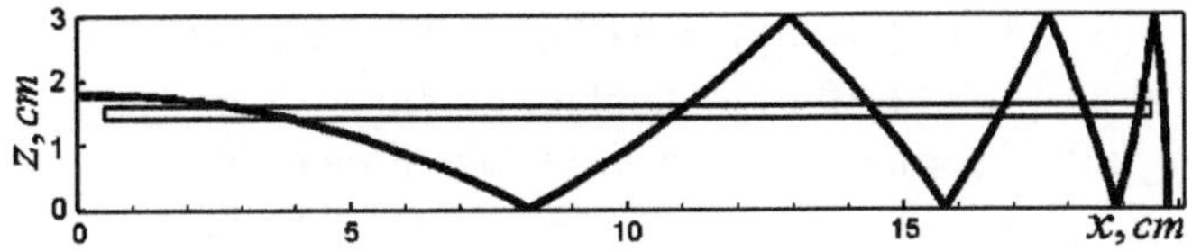


**Fig. 8.4.** A wave trace through the slit, $px = 0$, $pz = 1.5$ cm, $t = 6.7$

The slit scan imitates the motion of that point where the wave front and the slit intersect. Such a moving point leaves on the picture of the slit scan a track that looks like an inclined trajectory. The trajectory slope angle is determined by the scanning rate which is usually constant during the scan, and by the intersection point velocity which can vary.

As seen in Fig. 8.4, the wave during the given time intersects the slit simultaneously at several points. Quantity of the intersections and motion velocity of each of the points along the slit depends on the number $n$ of the wave reflection. These intersections form a set of the trajectories, all of them being at different inclinations.

All the trajectory set is described by the multiple-valued function $x(t)$. Accordingly, the multiple-valued derivative $dx/dt$ describes the velocity of every moving intersection, i.e., the slope angle which, generally, is a function of time. The diagram in Fig. 8.4 allows us to evaluate both the function $x(t)$ and the derivative $dx/dt$. Program 8.2 serves this purpose.

**Program 8.2** Linear scanning of a spherical wave in a pipe plane $y = 0$

```
x=t; x1=√(t²-0.25); x2=√(t²-9×0.25); x3=√(t²-25×0.25);
    x4=√(t²-49×0.25); x5=√(t²-81×0.25);
dx=Dt[x,t]; dx1=Dt[x1,t]; dx2=Dt[x2,t]; dx3=Dt[x3,t];
    dx4=Dt[x4,t]; dx5=Dt[x5,t];
gf=Plot[{x,x1,x2,x3,x4,x5},{t, 0, 5.5}, PlotPoints→100, Frame→True,
    PlotRange→{{-0.05,5.5},{-0.05,4.1}}, FrameLabel→{"t","x(t)"},
```

```
    PlotStyle→Thickness[0.008], DisplayFunction→Identity];
gf_txt=Show[Graphics[{Text["c",{1.1,3.2}.{0,0}],
    Text["n=1",{1.1, 0.45},{0.0}], Text["n=2",{1.95, 0.45},{0.0}],
    Text["n=3",{2.95, 0.45},{0.0}], Text["n=4",{3.9, 0.45},{0.0}],
    Text["n=5",{4.9, 0.45},{0.0}], DisplayFunction→Identity];
gdf=Plot[{dx,dx1,dx2,dx3,dx4,dx5},{t, 0, 5.5}, Frame→True,
    PlotPoints→100, PlotRange→{{-0.05,5.5},{-0.05,4.1},
    FrameLabel→{"t,cm","dx/dt"}, PlotStyle→Thickness[0.008],
    DisplayFunction→Identity];
gdf_txt=Show[Graphics[{Text["c",{0.3,1.3}.{0,0}],
    Text["n=1",{0.9, 3.6},{0.0}], Text["n=2",{1.95, 3.6},{0.0}],
    Text["n=3",{3, 3.6},{0.0}], Text["n=4",{4, 3.6},{0.0}],
    Text["n=5",{5, 3.6},{0.0}], DisplayFunction→Identity];
g1=[{gf,gf_txt}, DisplayFunction→$DisplayFunction];
g2=[{ gdf, gdf_txt}, DisplayFunction→$DisplayFunction];
```

The trajectories xi and the velocities dxi for each of five intersection points are computed separately in the program to display the procedure. Numerical simulation of the graphs gf and gdf of the functions xi and dxi is accompanied by corresponding text inscriptions gf_txt and gdf_txt in the program. Operations g1 and g2 prepare data for the trajectories and velocities plotted in Fig. 8.5. As a whole, the lines represent quantitative data on the shapes and inclinations of the trajectories within the chosen time interval $t\in [0, 5.5]$.

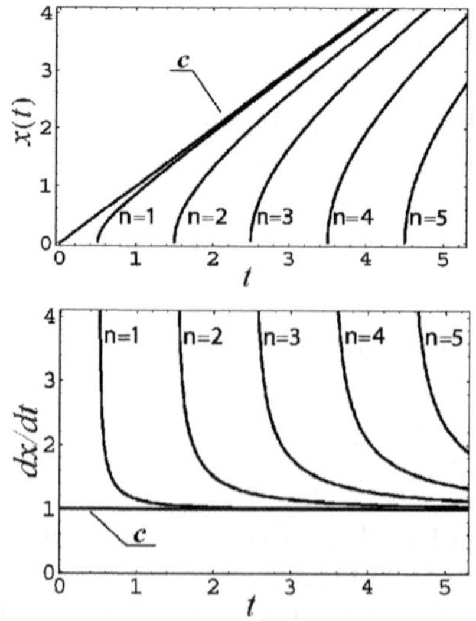


**Fig. 8.5.** Functions *x*(*t*) and *dx*/*dt* of slit scan for the wave n reflections

The straight line on the graph $x(t)$ passes through the origin of coordinates and corresponds to the plane wave propagation along the slot at the sound velocity $c$. The spherical wave originated on one of the horizontal walls comes into the slot field of vision at $t = 0.5$ for the first time ($n = 0$). From this time the first track on the slit scan appears, following the function $x(t)$. While the wave propagates, the track tends to the line related to the sound velocity, and merges finally with it at $t \approx 3$ [44].

The reflected waves intersect field of vision of the slot in regular intervals, $\delta t = 1$, and new tracks appear periodically on the graph $x(t)$, the period being equal to 1. Similar to the first track, the new tracks gradually come nearer to the sound velocity line.

The observer has customarily assumed that the velocity of the intersection point motion is the wave velocity. However the graph $x(t)$ in Fig. 8.5 shows that this opinion is inconsistent with experimental data, for this velocity comes as only an apparent one. Dependence of the apparent velocity on time and on reflection number n is shown by the computed derivative $dx/dt$ in Fig. 8.5. Here the horizontal straight line corresponds to the sound velocity. The remaining lines of the graph correspond to the motion velocity of the intersection points, or the apparent velocity of the wave after the reflection number $n = 1, 2, 3, 4, 5$.

There are two important circumstances that deserve attentive consideration. When the infinitesimal thin wave comes into contact with an infinitesimal thin slit, the apparent velocity would be infinitely large on the picture of the slit scan. However, if the wave and the slit are of finite width, the track of the intersection is diffused. Besides, starting and ending points of the track are not visualized in a proper way, for their images on the slit scan are diffused as well. The reason consists in the peculiarity of the Schlieren method registration employed in the measurement.

The Schlieren method visualizes optical perturbations by means of a so-called Foucault's knife. This technique is sensitive to optical density variation only in the direction normal to Foucault's knife blade. The blade has been oriented vertically in the experiments, as usual in such cases. The knife orientation means that the visualization efficiency is proportional not to density gradient $d\varrho/d\mathbf{r}$, but to its horizontal component,

$$\frac{\partial \varrho}{\partial x} = \cos\chi \left| \frac{d\varrho}{d\mathbf{r}} \right|. \tag{8.1}$$

Here $\chi$ is an angle between the slit and the normal to the wave front at the point of their intersection.

The angle of the tangency at the intersection instant amounts to $\chi = 90°$ ($\cos\chi = 0$), hence $\partial\varrho/\partial x = 0$, and the track of the wave at the onset turns out to be invisible. The track image becomes noticeable as soon as the value of $\cos\chi$ grows. At the same time, the wave amplitude damps in the course of propagation, $|d\varrho/dr| \sim 1/t$, and its component (8.1) tends to zero, $\partial\varrho/\partial x \to 0$. Therefore, the track on the slit scan disappears again at some instant. If the primary perturbation diameter is $2a$ when $t = 0$, then the component (8.1) can be written in the form

$$\frac{\partial \varrho}{\partial x} = \frac{\cos\chi\,|d\varrho/d\boldsymbol{r}|}{(a+t)}. \tag{8.2}$$

The function $\cos\chi$ in formula (8.2) for the slit field of vision can be expressed in terms of non-dimensional time $t$ to obtain an explicit form of the track's distinguishableness K [35]:

$$\mathrm{K} = \frac{a}{2\,t\,(t+a)}\sqrt{4t^2 - 1}. \tag{8.3}$$

If the characteristic dimension of the perturbation produced by the discharge is 0.2 cm, the non-dimensional size $a = 0.07$, just as in the experiment, this criterion can be represented by the ratio $\mathrm{K}(t)/\mathrm{K}_m$, where $\mathrm{K}_m$ is the distinguishableness factor maximum at the instant $t_m$. The factor as a function of time is plotted in Fig. 8.6.

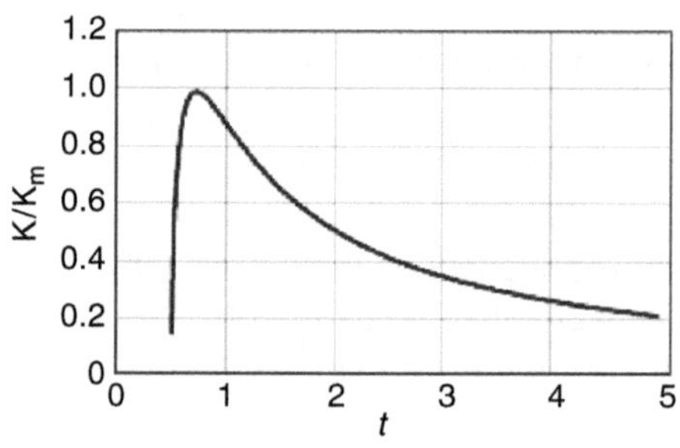


**Fig. 8.6.** Relative criterion of the track distinguishableness

The intersection point at the track end in the region $t < t_m$ moves first at velocity about 700 m/s, and when $t$ diminishes, the function $\mathrm{K}(t)$ value tends quickly to zero. The apparent velocity on the opposite wing of the function, where $t > t_m$, tends to ~400 m/s, and the function $\mathrm{K}(t)$ decreases slowly. Surely, the track on a Schlieren-gram is discernible provided that the factor K value exceeds some level $\mathrm{K}_{min}$, i.e., the condition $\mathrm{K} \geq \mathrm{K}_{min}$ should be satisfied.

As follows from the figure, this criterion has a maximum located at the argument $t_m \approx 0.75$, and the half-width is $\delta t \approx 1.6$. An opportunity to discern the track on the Schlieren-gram occurs during the time interval which approximates to $\Delta t \approx 4$. According to the function $x(t)$ in Fig. 8.5, this interval corresponds to the tracks by a length up to 12 cm, as it is observed in the slit scan process in Fig. 8.2.

So, the parameters derived from Figs. 8.5 and 8.6 that have been computed for the simple diagram drawn in Fig. 8.4, and the ones measured by the slit scan photo in Fig. 8.2 coincide with good accuracy. Notice, that the photos *c* in Fig. 8.3 confirm the fact that the secondary waves create a dense structure of perturbations in the flow just before the flame front. The perturbation structure is completely reproduced by the flame structure.

The frames *c* in Fig. 8.3 imply some important information on the process of the flow structure formation. These frames make it clear that the shape of the flame enveloping surface and, hence, the velocity of the flame propagation has maxima near the walls at the distance $\delta = 1.5–2$ mm. Hence distribution of the parameter oscillations is not uniform over the flow cross-section, and the oscillations are most intensive in the indicated area.

We shall return below to this problem which will be considered in detail. Here we shall mention only briefly the problem of the wall roughness which can happen to be the source of the oscillation under consideration.

It is commonplace to mention that roughness of the walls hastens sharply the turbulence development. A pipe wall being smooth is used to consider, if the ratio of a surface asperity $\delta z$ to the length $\Delta x$ of the boundary layer, say, at Reynolds number Re $\sim 5 \times 10^5$, does not exceed the value of $\delta z/\Delta x = 5 \times 10^{-4}$ [8]. The length $\Delta x$ in our case may by no means exceed the simple wave length; let it be $l \sim 10$ cm. Then the surface asperity initiates the flow perturbation provided that $\delta > 0.05$ mm.

Some special experiments have been performed to study whether the roughness is the source of the perturbations observed in our experiments, or the influence does not matter. Recall that the pipe walls have been polished. Some small threshold has been created in the pipe of square cross-section to test it. The threshold has been made by upward displacement of the pipe part following the visual section for the permissible asperity $\delta$ multiplied by 5, $5\delta = 0.25$ mm.

The frames *d* in Fig. 8.3 represent results of these experiments. As we see from the frames, the simple wave propagation near the asperity is accompanied by generating a sequence of waves, moving evidently all over the flow, including the direction against the current. The waves follow one after another with an interval of about 1 mm. This empirical observation implies that the simple wave structure is of the above mentioned discrete nature. In addition, the accurate concentric shape of the waves points to the threshold as a center of high pressure, which proves to be the source emitting secondary compression waves of rather considerable amplitude.

With the results described, we may conclude that the strengthening of the flow perturbations caused by the wall roughness can be easily explained by the wave mechanism of the turbulent oscillations. On the whole, analysis of the experimental data shows that the simple wave has a pressure discrete distribution including the compression waves maintaining the flow in front of the flame-piston. The waves are of a form close to spherical.

The flow perturbations originate at the channel walls as centers of local overpressure within the boundary layer, and propagate over the flow in the form of the spherical wave packets of thickness equal to the primary perturbation scale $2a$. Superposition of the packets forms the spatio-temporal field of the flow parameter oscillations. The assumption about pressure constancy in the flow cross-section, $\partial P/\partial z = 0$, can be accepted as the first approximation and only concerning the flow parameters averaged in time.

## 8.2 Boundary layer separation and flow perturbations

When a plane shock wave is reflected from a solid wall, the pressure $P_2$ behind the reflected wave exceeds the pressure $P_1$ behind the incident wave. The ratio of the overpressures $(P_2 - P_0)/(P_1 - P_0)$ in a polytropic gas with the exponent $\gamma$ is defined by the relation [56]:

$$\frac{P_2 - P_0}{P_1 - P_0} = 1 + \frac{1+(\gamma-1)/(\gamma+1)}{P_0/P_1+(\gamma-1)/(\gamma+1)}. \tag{8.4}$$

The ratio of the overpressures for an intense shock wave depends on the exponent $\gamma$, as seen from the following data:

$$\frac{P_2 - P_0}{P_1 - P_0} = 1+\frac{(\gamma+1)}{(\gamma-1)} = \begin{cases} 8 \text{ for } \gamma = 1.4, \\ 13 \text{ for } \gamma = 1.2, \\ 23 \text{ for } \gamma = 1.1. \end{cases} \tag{8.5}$$

In case of a weak shock wave, as well as a plane acoustic wave, this ratio of overpressures is reduced,

$$\frac{P_2 - P_0}{P_1 - P_0} \approx 2. \tag{8.6}$$

Presence of the boundary layer at the wall, the pressure profile and front shape of the wave introduce their specific contributions to the reflection process. Obviously, all these factors play a significant part, when the matter concerns the reflection of the spherical wave packet from a wall with a boundary layer over that.

Unfortunately this question has not been investigated experimentally. In the absence of direct experimental data relating to the reflection at conditions under consideration, there is no choice but to turn to the data of experiments on reflection of an oblique shock wave in a supersonic flow [8].

Figure 8.7 demonstrates the pressure distribution measured in a vicinity of the line where the oblique shock wave is reflected from a flat plate [57]. The plate is situated in an air flow at the following parameters: $M = 1.43$, $Re_x = 9 \times 10^5$, and the thickness of the laminar boundary layer 0.7 mm.

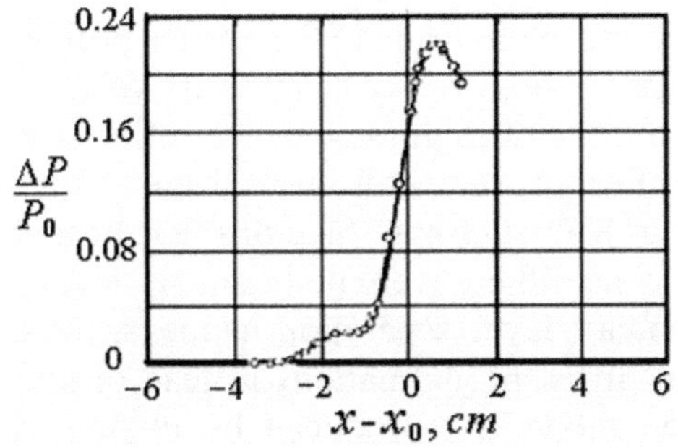


**Fig. 8.7.** Pressure at oblique shock wave reflection

Note at once, that according to [57] and [58], the pressure round about the reflection line parallel to the wedge blade grows irrespective of whether the boundary layer is separated or is not. At the same time, the boundary layer separation changes the structure of the reflection region and the reflection mechanism, as it follows from comparison of diagrams *a* and *b* in Fig. 8.8 [58].

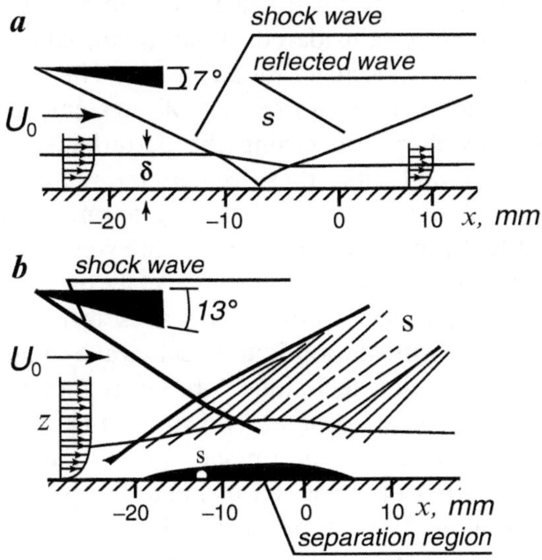


**Fig. 8.8.** Oblique shock wave reflection from a flat plate

The oblique shock wave in the experiments [58] is reflected from a flat plate situated in air moving at velocity $M = 3$, the boundary layer thickness being $\delta = 3$ mm. The oblique shock wave is produced by means of the wedge with an acute angle φ. The wave intensity depends on the angle and increases with angle growth.

At the angle value $\varphi = 9°$ the pressure ratio attains the level $P/P_0 = 2$, and there occurs the phenomenon known as a boundary layer separation. While the pressure nearby the reflection line exceeds 9°, but remains near the value, the boundary layer is reestablished behind the reflection region. But the angle's further growth leads to the layer's prolonged destruction.

While the shock wave is of weak intensity (e.g., at $\varphi = 7°$), and the boundary layer separation does not occur, the reflected wave leaves a trace on the photo in the form of one well-marked line. This situation is seen in the diagram *a* in Fig. 8.8. Then the trace of the reflected wave looks like a usual reflected wave, as if there were no boundary layer.

When the boundary layer separation happens, there appears a system of reflected lines of different inclinations instead of one trace. The pattern of such a reflection mode is reproduced by photo *b* in Fig. 8.8, where $\varphi = 13°$. The structure formation of those reflected lines is interpreted in reference [8] as follows.

The reflection of the intense shock wave entails appearance of the secondary waves at compression of various strength, and rarefaction waves. These propagate at different velocities, and hence the lines have different inclinations. As to the extension of the reflection pressure region, it is explained by the "pressure diffusion" during the boundary layer separation. However, there may exist quite another interpretation of the phenomenon.

According to the Huygens–Fresnel principle, every element of the separation region can be considered as the perturbation center, or source emitting an elementary spherical pressure wave. The real wave front in the whole is formed as a result of interference of the elementary waves radiated by the centers. However, concerning the perturbations in the separation region, we can think that, first, these centers are not coherent. Second, the waves emitted by various centers of the region are of different intensity. And last but not least, the flow carries away the waves emitted, and distorts the wave fronts in their ways.

Generally, the spherical elementary waves form the total cylindrical wave, when we deal with the oblique shock wave created by the wedge. The Schlieren visualization technique forms a two-dimensional picture in the direction of the wave axis, as shown in Fig. 8.8. Hence we may consider the wave propagation in a plane. Taking into account this notion and the three above reasons, we can estimate the process of the reflected wave propagation in the flow, as applied to the frame *b* in Fig. 8.8.

Motion of the wave probe point depends on its initial position defined by the angle $\chi \in [0, \pi]$, on the wave velocity $u/c = m$ ($m \geq 1$), and on the flow velocity $U/c = M$. Naturally, the flow drifts the wave and distorts its front shape as well as the probe point trajectory. The angle α between the trajectory and $x$-axis is determined by ratio of the point transverse velocity $m\sin\chi$ to the longitudinal velocity $M+m\cos\chi$:

$$\mathrm{tg}\alpha = \frac{m\sin\chi}{M+m\cos\chi}. \tag{8.7}$$

The function (8.7) for given $m$ and $M$ has a maximum $\alpha_m$ at some values of the argument $\chi_m$. The maximum position on the axis $\chi$ can be found by equating the derivative from the function $\mathrm{tg}\alpha$ with respect to the angle $\chi$ to zero,

$$\frac{m\cos\chi}{M+m\cos\chi}+\frac{m^2\sin^2\chi}{(M+m\cos\chi)^2}=0. \tag{8.8}$$

Solving the equation for the value of $\alpha_m$ known from experiment results, e.g., from graph *b* in Fig. 8.8, we obtain equations that determine the wave velocity $m$ and the angle $\chi_m$:

$$\left.\begin{aligned} m &= M\sin\alpha_m, \\ \chi_m &= \pi-\arccos\left(-{}^{m}\!/_{M}\right). \end{aligned}\right\} \tag{8.9}$$

The diagram *b* in Fig. 8.8 shows that the angle $\alpha$ has upper limit $\alpha_1$. Also, the angle $\alpha$ has some lower limit $\alpha_2$. It follows from formula (5.15), according which the wave front does not exist as a whole beginning from some direction $\chi_2$, whence $\chi < \chi_2$. Therefore, the real values of $\alpha$ belong to the interval $\alpha \in [\alpha_1, \alpha_2]$.

Among the set of reflected waves, those that have the greatest intensity may be expected to leave some noticeable marks on the Schlieren slit scan along their traces. For example, a wave of great intensity can be recognized by indirect signs, such as some bulging of the boundary layer. A similar effect is observed on the diagram *b* in Fig. 8.8 within the sector S with the reflection point s as a center. Measurements of the angle value $\alpha_1$ in the sector result in $\alpha_1 \approx 39°$. Then it follows from relations (8.9) that $m \approx 1.9$ and $\chi_1 \approx 141°$.

When $m$, $M$ and $\chi$ (or $\vartheta$) quantities are known, equations (5.16) and Program 5.1 give an opportunity to compute the trajectory of any probe point. The wave front of radius $\ell$ is clear to represent in essence the locus of the tips of the trajectories at the same length $\ell$. The wave front distortion may be evaluated in the flow with non-uniform velocity distribution by means of computation of these trajectories. The data concerning the probe point trajectories and the wave front configuration constitute a new model of the wave reflection.

The diagram in Fig. 8.9 illustrates the model of the reflection process shown on frame *b* in Fig. 8.8. The slant lines in the figure imitate the trajectories of seven probe points within sector S, which have the common center s. The heavy line within the sector depicts the configuration of the originally spherical wave front, $\chi \in [0, \pi]$, deformed as a result of its drift by the flow while propagating.

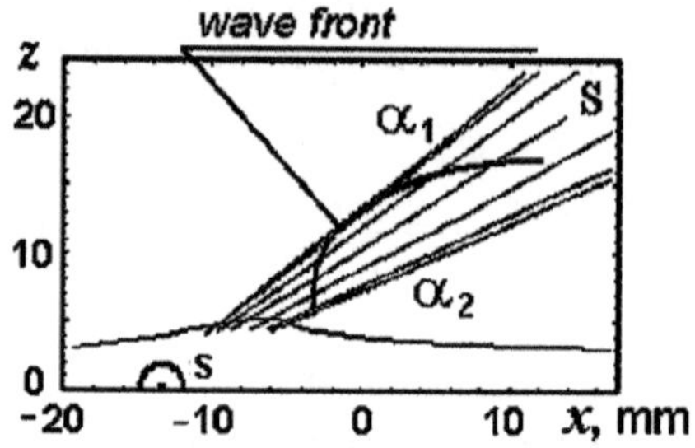


**Fig. 8.9.** Model of wave reflection shown by diagram *b* in Fig. 8.8

Properties of the slant lines (Fig. 8.9) obtained from the numerical simulation practically coincide with the lines within the sector S observed in the experiment (frame *b* in Fig. 8.8). This coincidence serves as an experimental verification of the suggested interpretation of the wave reflection at moderate intensity. This interpretation can be applied to all the slant lines in Fig. 8.8*b*, with the exception of that line which corresponds to propagation of the very early perturbation.

The slope angle α of the line changes quickly. The angle in the beginning makes up ~35°, but soon decreases down to ~26°, that means a motion deceleration from $m \approx 1.7$ to $m \approx 1.3$. Generally, such quick deceleration of the wave can be explained, only if the wave is of spherical, but not of cylindrical symmetry. The source of the spherical wave instead of the expected cylindrical one might arise as a protuberance, or even several protuberances on the leading edge of the wave reflection region. Thus, the offered concept of the wave reflection does not contradict the empirical observations which could be interpreted in all points.

Surely the oblique shock wave reflection from a wall with boundary layer, that have just been considered, somewhat differs from the reflection of the spherical acoustic wave packet. Indeed, the oblique shock wave is formed in a supersonic stream, and the region of its reflection comprises rather strong perturbation which is stationary and, besides, is of cylindrical symmetry. However the cited data serve as an indirect proof of the fact at least, that a local perturbation arising at the boundary layer separation propagates in a flow as a wave with a central symmetry.

## 8.3 Distribution of oscillations in flow cross-section

The structure of the perturbations and spatio-temporal field of the oscillations have been computed in Chapters 5 and 6 for the main potential part of the flow. The pressure gradient $\partial P/\partial z$ in this part of the flow is small, as demonstrated in Fig. 3.1, and the trajectories of the wave probe points are described by rather simple equations (6.1), (6.2).

The straightforwardness of the trajectories means that the direct and reflected waves take part in the superposition at any point of the flow. The only impediment for estimation of a wave contribution to a parameter perturbation at the observation point is the wave damping, i.e., the distance $\ell$ from the place of its arising to the superposition scene.

Early in Chapter 5, it has been mentioned that the intensity distribution of the parameter oscillations does not keep a constant level over the flow cross-section. Also, it has been mentioned that the oscillations are most intensive near the wall. These features of the oscillation intensity distribution inhere in the flow within the simple wave as well. As emphasized in ([3], §42), the fluctuating flow near the wall "has some peculiar features that have not yet been given an adequate theoretical explanation." However the peculiarity of the oscillations (fluctuations in terms of [3]) can be interpreted in the frame of the wave conception of the turbulence.

As it follows from equation (5.14), the velocity complex profile in a boundary layer makes considerable modifications in the character of the wave propagation. Certainly, some spatio-temporal field of hydrodynamic parameter oscillations is formed in the boundary layer, and such a phenomenon as wave superposition exists. However the wave vectors **s** of the wave taking part in the superposition alter very much the directions while propagating through the boundary layer.

Accordingly, the wave contribution to the parameter perturbations at the chosen observation point happens to depend on a variety of factors in a not quite evident way. One of the factors is whether a certain perturbation arises upstream or downstream concerning the observation point. Then, what is the distance between the wave origin spot and the observation point. At last, what is the altitude of the observation point above the wall.

The trajectory description technique has been expounded in Chapter 5. The equation set (5.14) relates to general case of wave propagation. Now let us consider a solution of the equations for the plane $y = 0$. The set (5.16) assigned for the plane $y = 0$ implies that the trajectories start at the origin $x = z = px = pz = 0$. However if we deal with wave superposition, the parameters $px, pz \neq 0$ should be taken into account.

We have already known, that the trajectory maximum $z^*$, if any, depends on the angle value $\vartheta_0 = \vartheta^*$. Accordingly, if such maxima are located in one cross-section of the flow, say at $x = 0$, the configurations of the trajectories in the plane $y = 0$ look like as they are shown in Fig. 8.10.

This graph represents the fluid motion in the flat channel at the flow parameters $m = 5$ and $M_0 = 0.05$. The trajectories are arranged in such a way that the maxima are $z^* = \{0.075, 0.186, 0.341, 0.422, 0.463, 0.484\}$. The maxima arrangement requires the angles $\vartheta_0$ and the origin places $xp^*$ being $\vartheta^* = \{74, 72.7, 72, 71.85, 71.815, 71.807°\}$, $xp^* = -\{1.17, 2.79, 5.74, 8.5, 11.2, 14.3\}$. As a matter of fact the values $z^*$, $\vartheta^*$, $px^*$ are bound up with each other at given parameters $m$ and $M_0$, as the relation (5.17) testifies.

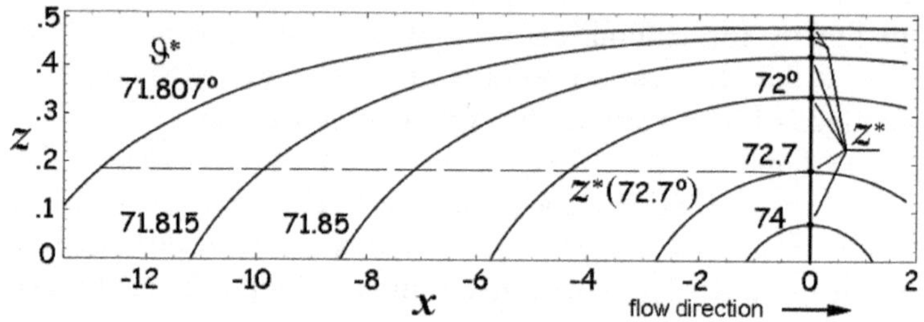


**Fig. 8.10.** Trajectories of probe points for the angle $\vartheta_0 = \vartheta^*$ values that correspond to maxima $z^*$ originated at the spots $xp^*$

The horizontal broken line drawn at the level $z^*(\vartheta^* = 72.7°)$ intersects all the trajectories thereof initial angle $\vartheta$ and the distance $xp$ are less than $\vartheta^* = 72.7°$ and $xp(72.7°)$. It follows that a set of the trajectories with many other various values of $xp$ and $\vartheta$ may pass through the point $z^*(\vartheta^* = 72.7°)$. However there is a limiting distance $xp^*$ beyond which a trajectory ascending branch does not reach the point. Such limiting trajectories are defined by the initial angles $\vartheta^*$ which depend on the flow parameters $m$ and $M_0$. It is those trajectories that are shown in Fig. 8.10.

Now there is a need to elucidate, which of the trajectories reach some arbitrary observation spot at the coordinates $\{x_0, z_0\}$. The equation (5.18) implies and the example in Fig. 8.10 illustrates, that some perturbation gets into the observation spot by both ascending and descending branches of the trajectory. To obtain more thorough information on the subject of which trajectory may reach the point $\{x_0, z_0\}$, it is better to take advantage of the simple flow with well-known parameters, e.g., $m = 5$, $M_0 = 0.05$ at $y = 0$.

The wave centers situated upstream and downstream from the point will be denoted as $xp^- < x_0$ and $xp^+ > x_0$, accordingly. As equation (5.18) and Program 5.1 predict, the perturbation from the region $xp^+$ hits the observation spot $\{x_0, z_0\}$ via the ascending branch, whereas both branches, ascending and descending, can serve the purpose, when the perturbation starts in the region $xp^-$. To evaluate the effects of the perturbations on the superposition of the oscillations, transmitted along the trajectories of these two types, we will compare this effect in two spots.

Let one of them be near the wall $\{x_0 = 0, z_0 = 0.1\}$, and the other take its place in the middle of the channel $\{x_0 = 0, z_0 = 0.5\}$. The trajectories of different perturbation waves passing through these two spots are plotted in Fig. 8.11. The upper graph represents the trajectory set arriving at a spot remote from the wall $\{x_0 = 0, z_0 = 0.5\}$. The lower graph depicts the trajectories around a spot lying near the wall $\{x_0 =0, z_0 = 0.1\}$. Only the waves originated on the lower wall are shown in the figure for the purpose of the graph simplicity.

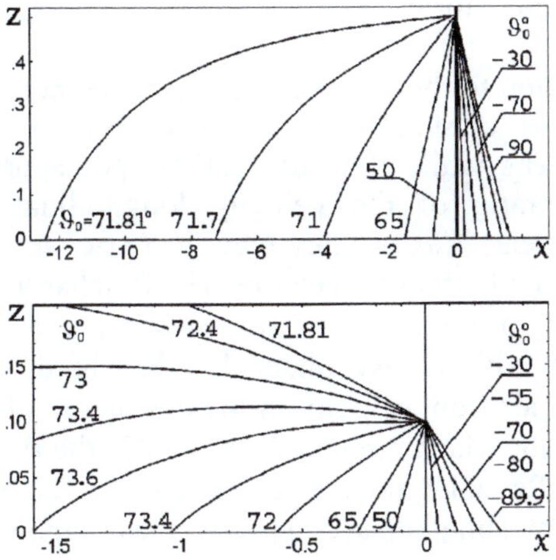


**Fig. 8.11.** Trajectories directed to observation spots $\{x_0, z_0\} = \{0, 0.5\}$ and $\{x_0, z_0\} = \{0, 0.1\}$

Looking at the graphs, we conclude that the perturbations coming from the region $xp^-$ reinforce the longitudinal component $u$ of the velocity oscillations, and the perturbations from the region $xp^+$ promote intensification of transverse component $w$ of the oscillations. Also, we can't help paying attention to the region $xp^-$ being much longer than the $xp^+$ region. Therefore, the oscillations directed along the stream get some advantage, all other things being equal.

Now we are about to investigate how total orientation of the oscillations created by the perturbations depends on the distance $z_0$ of the observation spot from the wall, or what is the distribution of the velocity u components along $u$ and across $w$ of the stream in the flow cross-section [33].

Initial orientation $\vartheta_0$ of the trajectories that come to any of the two observation spots from the region $xp^+$ varies in rather broad limits which are $\vartheta_0 \in [-\pi/2, 0]$, to be exact. Hence the contribution made by these waves to the parameter perturbations at both spots $z_0 = 0.5$ and $z_0 = 0.1$ do not differ noticeably. The essential difference in the contributions appears when the waves come from the region $xp^-$.

The trajectory's initial orientation, that comes to the observation spot from the region $xp^{-}$, lies within some narrow interval, for the angles $\vartheta_0$ are restricted by limiting angle $\vartheta^*$, $\vartheta_0 \in [\sim\vartheta^*, \pi/2]$. The limiting angle for the observation spot $z_0 = 0.5$ amounts to approximately $\vartheta^* = 71.8°$. So that the spot is accessible only for the ascending branch of the trajectories. The similar limiting angle $\vartheta^*$ exists for a spot situated near the wall too, in particular if $z_0 = 0.1$, then $\vartheta^* = 73.6°$. However, in contrast to the previous case, the lower spot is accessible both for the ascending and descending branches of the trajectories, and the latter maxima belong to the range $z^* \in [0.1, 0.5]$.

The contribution to the oscillations due to the perturbations coming by such trajectories grows along with the length of the region $xp^{-}$. In its turn, the region lengthens while the observation spot approaches to the wall. Accordingly the ratio of the velocity longitudinal component to the transverse component, $u/w$, progressively increases.

Just such a trend was observed by H. Reichardt in experiments [59] with the channel of rectangular cross-section, and in P.S. Klebanov's experiments [60] with a flat plate. H. Reichardt measured turbulent oscillations of the air flow velocity in the channel by the width of 1 m and height 24.4 cm at the Mach number $M_0 = 0.003$ (the flow velocity $U_0 = 100$ cm/c), when the Reynolds number exceeded $10^4$. Distributions of the root-mean-square velocity longitudinal and transverse components across the flow,

$$\sqrt{\overline{u^2}}\Big/\left(\sqrt{\overline{u^2}}\right)_{\max} \text{ and } \sqrt{\overline{w^2}}\Big/\left(\sqrt{\overline{u^2}}\right)_{\max}, \tag{8.10}$$

are reproduced in Fig. 8.12 by points. As the distributions are symmetric concerning the channel middle, a half of the curves is shown on the graph.

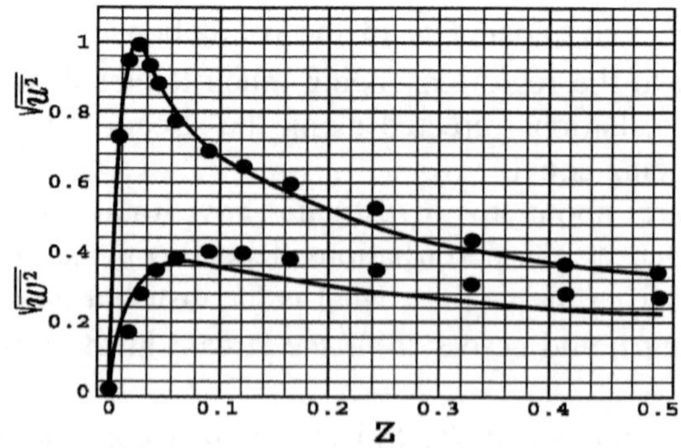


**Fig. 8.12.** Longitudinal $\sqrt{\overline{u^2}}$ and transverse $\sqrt{\overline{w^2}}$ components of the velocity oscillations in relative units

It is evident from H. Reichardt's measurements, that the transverse component of the velocity oscillations displays feebly marked dependence on the coordinate $z$, whereas the longitudinal component has the explicit well-defined maximum near the wall, at $z \approx \delta \approx 0.03$. This result, which has had no consistent explanation until recent time, is naturally interpreted on the basis of the stated understanding of the turbulent fluctuations as the oscillations of acoustic nature.

Moreover, there appears an opportunity to obtain a quantitative description of the oscillation distribution observed in the flow cross-section within the frame of this acoustic approach. Employing the relation (5.6) allows us to formulate the components of the velocity oscillations in the form:

$$\left.\begin{aligned} u &\sim f\left(\frac{\ell_n-\ell}{a}\right)\sin(\vartheta)/\ell, \\ w &\sim f\left(\frac{\ell_n-\ell}{a}\right)\cos(\vartheta)/\ell. \end{aligned}\right\} \tag{8.11}$$

Here $\ell$ is the current distance covered by the wave front, $\ell_n$ denotes the path length from the perturbation origin to the observation spot $\{x_0, z_0\}$, angle $\vartheta$ implies the slope of the trajectory at the point $\{x_0, z_0\}$, and $a = \delta$ means the characteristic dimension of the primary perturbation, being about the thickness of the boundary layer.

Generally, the quantities of the perturbation velocity components, as well as the outcome of the superposition, are determined by the relations (5.14). However it is reasonable to simplify the calculation technique for the sake of better comprehension of a physical picture of the process and the space saving. For that we replace the curvilinear trajectory between the perturbation centre $\{x = xp,\ z = 0\}$ and the observation spot $\{x = x_0 = 0,\ z = z_0\}$ by a straight line, as shown in Fig. 8.13. The straight line length does not practically differ from the path length $\ell$ along the real trajectory.

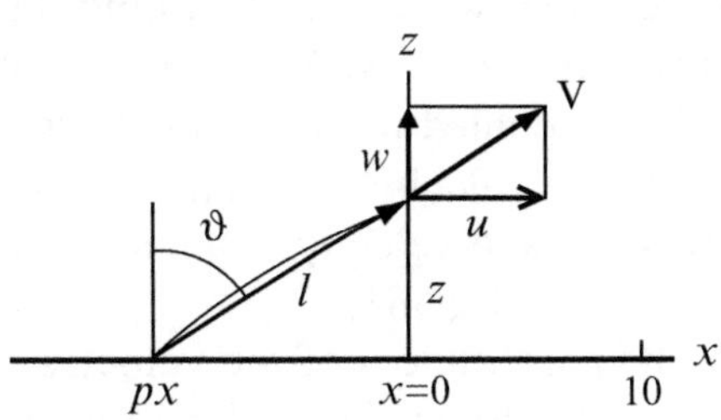


**Fig. 8.13.** Design of bowed trajectory replacement by straight line

Now let us assume that the parameter $xp$ distribution along the $x$-axis is uniform, and take into account only those perturbations which arise in the range $xp \in [-10, 10]$. The wave amplitude outside of the range is too small to be included in the calculation as a first approximation. Then the root-mean-square velocity components of the oscillations can be expressed through the arguments $xp$ and $z_0$ as follows:

$$\left.\begin{aligned} \sqrt{\overline{u^2}} &\sim \sqrt{2\int_a^{10}\left(\frac{(1+\varepsilon)\,xp}{xp^2+z_0^2}\right)^2 d\,(xp)}\,, \\ \sqrt{\overline{w^2}} &\sim \sqrt{2\int_a^{10}\left(\frac{(1-\varepsilon)\,xp}{xp^2+z_0^2}\right)^2 d\,(xp)}\,. \end{aligned}\right\} \tag{8.12}$$

The factor ε is inserted into integrands to compensate the fault in the contribution into the oscillations made by the perturbations, resulting from the replacement of the real trajectory by a rectilinear path. The mere replacement would shorten a little the longitudinal component of the velocity oscillations, and lengthen the transverse component. The value $\varepsilon = 0.2$ has been used in the computation. Also the lower limit of integration, *a*, has been included to take into consideration the fact, that an outer problem of the wave propagation is solved in the field $r > a$. Hence the integration is carried out outside of the boundary layer, $z_0 > a \approx \delta$.

Thus the outer problem of the wave field (2.30) determination is reduced to solution of equations of the type (5.2). The inner problem solution, when integration (2.30) is carried out at $r < a$, leads to the components of the velocity oscillations changing from their maximum values at the edges of the primary perturbation to zero on the wall surface, $z_0 = 0$.

The components of the velocity oscillations, computed according to formulas (8.12) with their definition within the boundary layer, are presented on the graph in Fig. 8.12 by solid curves along with dots for the experimental data. Comparison of the experimental and calculated data shows that the simulated functions and experimental dependences are well matched [33].

H. Reichardt's thorough measurements give us a chance to evaluate parameters of that acoustic wave which would be capable to create observable oscillations of the velocity. According to data of the experiment, the flow mean velocity amounted to $U = 100$ cm/s, and the level of the effective amplitude of the velocity oscillations made up u ~ 5 cm/s. Oscillations of the pressure and velocity in an acoustic wave are bound by the simple relation $p/P_0 = \mathrm{u}/c$ [3, §64].

Hence, knowing the amplitude u of the velocity oscillations from the experiment [59] allows us to evaluate the effective amplitude of the pressure oscillations as $p/P_0 \approx 10^{-4}$. For example, in air at the atmospheric pressure ($p_0 \sim 10^5$ Pa) this amplitude will be $p \approx 10$ Pa, or ~0.1 mmHg. As a matter of fact this experimental quantity corresponds to the level of loudness of human speech.

Thus, the acoustic model of turbulent oscillations (usually called fluctuations) gets experimental verification in all the basic theses.

# 9
# Transition from normal combustion to detonation

## 9.1 Short history of the problem

Research on non-stationary combustion and detonation origins began after catastrophic explosions occurred in collieries of France and Belgium. Subsequent study of the problem was also stimulated by the phenomenon of a "knock" in internal-combustion engines and intensification of fuel burning in industrial fire-chambers and jet engines.

Detonation in natural explosive gas mixtures, following a thermal initiation of slow combustion, was discovered almost simultaneously by M. Berthelot, P. Vieille [61] and E. Mallard, H. Le Chatelier [62]. Then V.A. Michelson [63] partitioned the phenomenon into three transient stages depending on the flame propagation character. These stages are slow flame propagation, flame accelerated motion, and detonation.

V.A. Michelson paid attention to the important fact that combustion within a flame entailed a pressure increase in the flame's zone. This observation furnished a vital clue to understanding of the combustion altered regime at all three stages of flame propagation. Unfortunately, this observation and the idea itself were forgotten for some time.

The concept of an effect of acoustic waves on flame structure and propagation was first stated by H.B. Dixon [64]. He confirmed experimentally the guess surmised by H. Le Chatelier that some "invisible waves" spread quickly before the flame. Later on, the phenomenon of normal combustion (deflagration) transition into detonation was investigated by many authors, in particular in the works [49, 50, 65–68]. This list of references is compiled in chronological order, and certainly does not contain all the publications on this problem.

Combustion represents a process resulting in energy liberation in the zone of chemical reactions. Inasmuch as the combustion zone is spatially restricted, the energy release is accompanied with expansion of the combustion products to set the surrounding gas in motion. Consequently combustion is a hydrodynamic process as well.

When an explosive gas mixture is ignited at a closed end of a pipe, the combustion product expansion sets the fresh gas mixture in motion in front



L.N. Pyatnitsky, *Turbulence Nature and the Inverse Problem*, Fluid Mechanics and Its Applications 89, 125–154.

of the flame. The moving fresh gas is compressed beginning from the forward boundary of the motion, and the gas temperature, density and velocity increase as well as the pressure. In due course of time a shock wave is formed at the front boundary. Meanwhile, the gas pressure and density decrease in the combustion products behind the burning zone where the chemical reactions occur. All this results in formation of the so-called "double saltus" [50], i.e., a column of gas flowing between the running shock wave and a traveling front boundary of the combustion products.

The gas convective stream causes a flame surface increase and a flame acceleration that are followed by a mixture combustion intensification and the double saltus strengthening. Expansion of the combustion products acts as a piston moving with acceleration. As a consequence, the gas mixture motion within the column is quickened.

When the velocity reaches some great value, the flow and the flame become turbulent [49]. The velocity of a turbulent flame exceeds considerably the normal flame velocity, and the turbulization serves as a reason for the second flame acceleration.

The elementary compression waves, which are emitted by the accelerating flame-piston, overtake each other and produce a compression wave of considerable amplitude in the end [46]. The latter soon transforms to the front shock wave, where the temperature rises, and detonation occurs [49, 50, 65].

The exposition of the non-stationary transition into detonation has met with some formal difficulty. According to the theory of stationary propagation of the chemical reaction zone, whose foundation was laid by D. Chapman [69] and E. Jouguet [70], the conservation equations can be solved by using features of the well known Hugoniot adiabatic curve for detonations and deflagrations.

Two branches of the curve are of practical significance. One of them describes the reaction zone propagation at a velocity up to 100 m/s, and corresponds to normal combustion. The second branch is applicable to velocity over 1,500 m/s, and accounts for the detonation propagation. But the curve of Hugoniot does not interpret the transient regime of combustion. The theory of the Q-curve [71] was developed to make up for this deficiency, and to obviate the theory's formal difficulty [49, 72].

The established apprehension concerning the slow combustion transition to detonation sufficiently agrees with observational data. Nevertheless, one point is worthy of being noted. The authors of previous researches have proceeded from the assumption that all the processes develop in one dimension. Also such important factors as pressure increase in the zone of the energy release have not been taken into account. In particular, effect of the pressure waves caused by the energy release on the process development has not been considered at all.

Hence, understanding of the slow normal burning transition into the detonation needs some additional thorough consideration.

## 9.2 Exposition of flame propagation in a pipe

The study of the flame propagation process and transition of the normal flame into detonation was carried out in a pipe of circular cross-section, diameter of 3 cm, and in a pipe of square cross-section with the square side 2.9 cm. Each of the pipes had modular construction consisting of interchangeable sections, including its own optical module, the visual section, for the process development visualization and registration by means of Schlieren slit scan recording and Schlieren high-speed filming (see Chapter 8). An overall view of the visual modules is shown in Fig. 9.1. The overall length of each pipe was 155 cm.

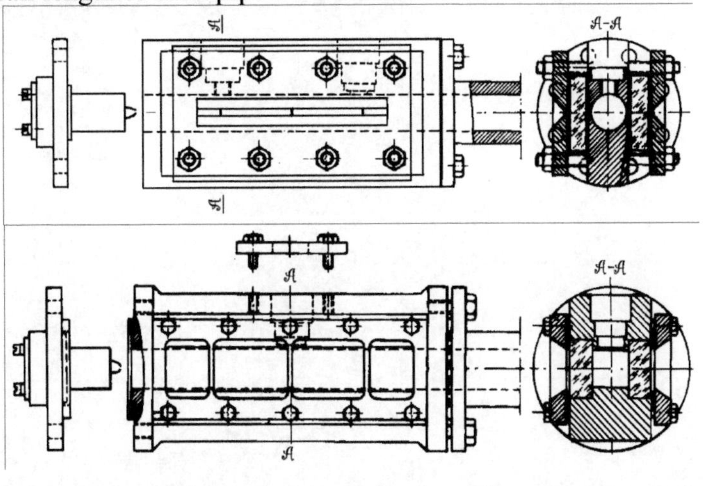

**Fig. 9.1.** Visual modules of pipes of circular and square cross-sections

The experimental facility provided the equipment for making the visualization of all kinds of perturbations in the form of moving optical non-homogeneities by the Schlieren and shadow methods. Also, the set-up included devices for visualized scene registration by slit scanning and filming, and continuous pressure recording with temporal resolution $10^{-4}$ s and spatial resolution about 1 cm.

An explosive mixture of methane with oxygen, $CH_4 + 4O_2$, was selected after the preliminary experiments. The quantity of molecules in such a mixture composition does not change in the course of an oxidation reaction. In addition, the combustion products comprised a rather small portion of the water molecules, condensation of which might complicate the interpretation of the observation results. The Schlieren slit scan photo of the flame propagation process in the pipe of the circular cross-section is presented in Fig. 9.2 [51].

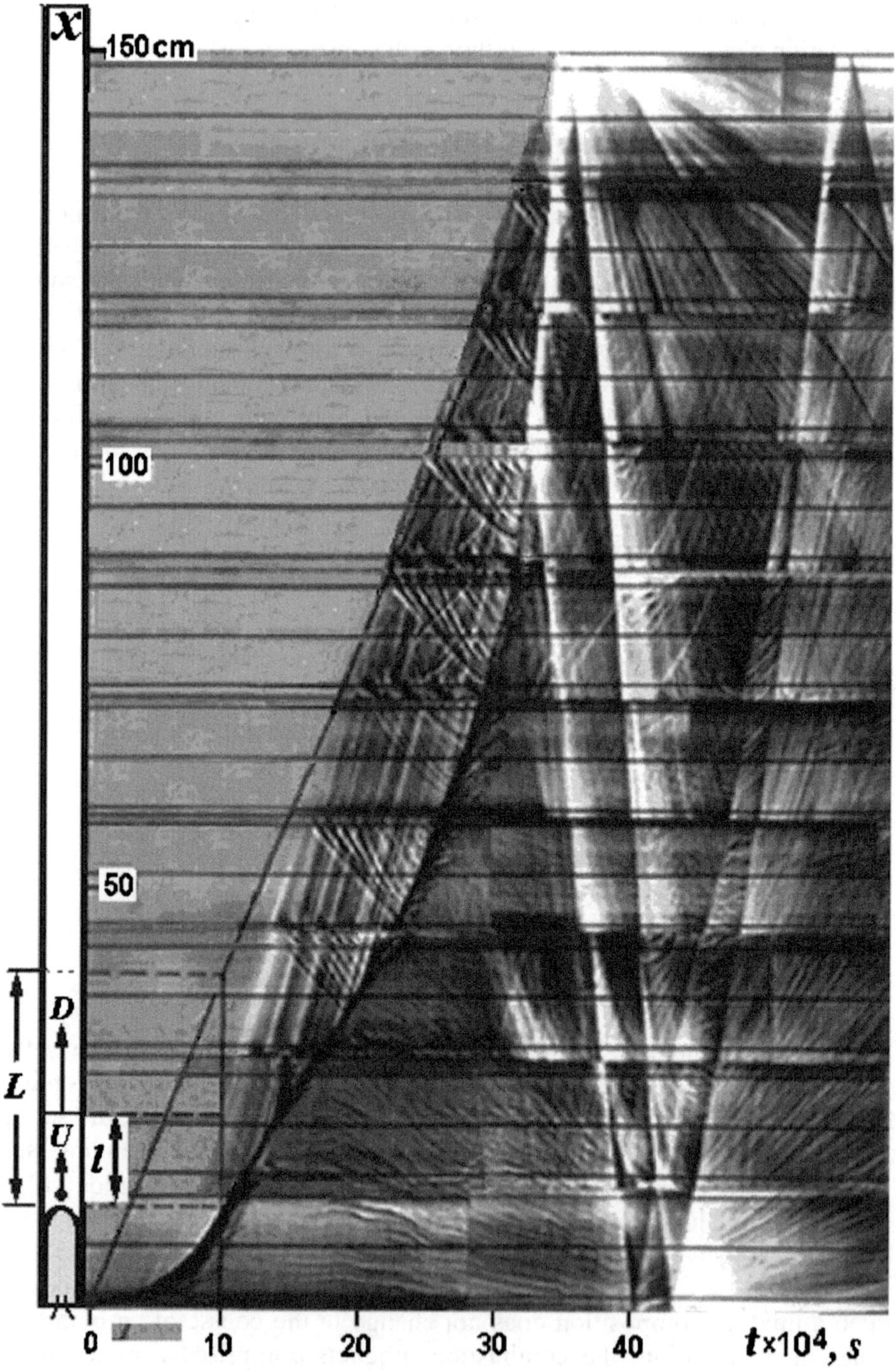


**Fig. 9.2.** Slit scanning of the flame propagation process in a pipe of circular cross-section

The transition to the detonation occurs at a distance about 110 cm from the mixture ignition site. Inasmuch as the pipe length amounts to 155 cm, the shock waves reflected from the opposite end of the pipe meet the detonation wave in plenty of time after the detonation has already occurred. In this sense the pipe can be considered as being infinitely long. The flame propagation is registered as the dependence of the coordinate $x$ on time $t$ beginning from the ignition instant. The process details are essential, and the time and space resolutions have been provided for.

The time resolution and tracing duration depend on the velocity of scanning, and are easily controlled, though the two parameters have mutually exclusive requirements. The visualized picture is recorded at a large scale to augment the space resolution whenever possible, but the proper field of vision defines the visual module dimensions. As a result, all the picture along the pipe length $L$ consists of ten parts implying ten successive positions of the visual module. Hence the photo in Fig. 9.2 is actually a mosaic made up of separate fragments obtained in different experiments.

Horizontal black strips are dinted by three opaque scale marks of the visual module. The moment and place of the mixture ignition on the photo are situated at the point $x = 0$, $t = 0$. At the photo's left has been drawn a schematic diagram of the double saltus instant configuration in a pipe. The diagram includes the flame sketch, the first intense shock wave at the distance $l$, and the first sound wave which is at distance $L$ from the mixture ignition spot. The diagram refers approximately to the time instant $t = 1$ ms.

Hence the region of the whole simple wave has length $L$, the real double saltus is of length $l$. The flame propagation velocity is denoted through $U$, the shock wave through $D$. The velocity of the gas flow $W$ and the flame normal velocity $u_0$ are not shown on the diagram. The velocity $u_0$ refers to flame's front motion relative to the fresh mixture in the direction normal to the flame surface.

The broken slant line reproduces the trace from the ignition site, which the first acoustic wave would leave on the linear slit scan, if the imaging method were more sensitive. The line indicates the forward edge of the double saltus, i.e., the front of the simple wave to be exact, however the first shock wave seems to be the real boundary. The broad dark track of alternating slope on the photo relates to the flame motion.

The numerous nearly straight lines outgoing from the flame, which propagate before the flame, are the compression waves. They originate in the combustion zone due to expansion of the combustion products. There stand out traces of the forward and backward detonation waves, and the powerful shock waves reflected from the pipe ends. The tortuous lines behind the flame are the traces of the non-homogeneities produced by motion of optical inhomogeneities in the combustion products. The slope of the traces corresponds to local velocity of the perturbations of various kinds in the hot gas.

We repeat once again, the Schlieren method enables us to distinguish a perturbation as an optical inhomogeneity in a gas under examination, provided that the inhomogeneity gradient is not outside the scope of the sensitivity. Generally the detection sensitivity depends on the gas temperature, and the temperature rise reduces the detection opportunity. However a strong perturbation with a heavy gradient is visible even in combustion products, despite the high temperature.

So, the higher the temperature of the gas, the more strong perturbation is required to be observed. The perturbation motion leaves on the slit scan the slant line. The field of the lines contains information on the propagation velocity (concerning the pipe) of the pressure waves, of the flame, of separate perturbations in the flow in front of the flame, and in the combustion products.

One more kind of traces needs some special explanation, for the traces are located within the double saltus. There are small cavities of 0.2 mm depth and 3 mm width at the junctions of the optical glass and the observation slot in the visual module body of the circular pipe (Fig. 9.1). As a matter of fact the cavity is nothing but a local roughness. The effect of the cavity is demonstrated by the photo in Fig. 9.2.

When a set of pressure waves emitted by the flame passes near the cavity, the appropriate sequence of the perturbations arises in the form of weak spherical pressure waves moving within the simple wave. The waves propagating in the slit's field of vision leave the traces on the slit scan image, looking like the curved lines that correspond to the motion against the current.

The visual module of the pipe of square cross-section has no such cavities, for the glass slab and metal wall have at the junctions nearly a single continuous whole surface. Therefore, the reflected waves do not arise. Analysis of the Schlieren photo of the slit scan in Fig. 9.2, and thereof comparison with the slit scan in the pipe of square cross-section shows, that the local perturbations do not make an impact on the explored process development. Nevertheless, the perturbations in the circular pipe lessen the path of the normal combustion transition into detonation for a small part of a millisecond, and reduce the length of the transition from 5 to 10 cm.

The velocity of any perturbation can be specified by measurement of the trace slope angle $\beta$ and the scale factor $k$ peculiar to the slit scan image. For instance, the flame velocity $U$ in Fig. 9.2 is determined as

$$U = k\mathrm{tg}\beta. \tag{9.1}$$

The flame velocity $U$ as a function of the distance $X$ covered by the flame is represented in Fig. 9.3. The slit scan photo in Fig. 9.2 has been used for the measurements. A fleeting glance at the process development allows us to appreciate the character of the transition to detonation, and to divide the process into its separate stages.

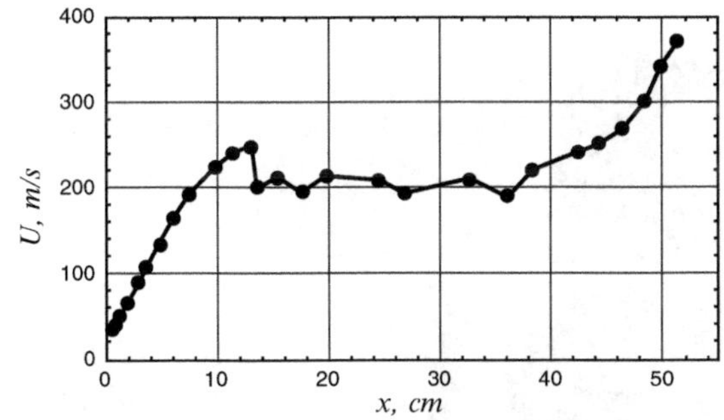


**Fig. 9.3.** Flame velocity along the trajectory as the slit scan shows

The flame's normal velocity in the mixture $CH_4 + 4O_2$ is equal to 2.3 m/s. Expansion of the combustion products enlarges the propagation velocity of the flame front along the pipe axis up to ~23 m/s. For the reasons expounded in Chapter 8, according to the slit scan, the flame starts motion immediately from the acceleration, and in ~1.3 ms after ignition the propagation velocity reaches the maximum value which amounts to ~250 m/s.

Then, during a relatively long period, ~2.7 ms, the flame moves almost at the constant velocity ~200 m/s, being accompanied by small oscillations about the mean value. After this quasi-stationary motion, the flame propagation mode is changed, and the second progressive acceleration is observed, which is followed by formation of the detonation.

Thus, the entire process of the normal combustion transition to the detonation lasts approximately 3.7 ms and consists of three stages, to say nothing about the detonation itself. The main transitional shapes and structures that the flame takes while propagating are assembled and represented by the frames of high-speed photography in Fig. 9.4. Now we will dwell upon each of these three stages.

## 9.3 Initial stage of the flame propagation

The graph in Fig. 9.3 testifies that the initial stage of the flame propagation process begins naturally from the explosive mixture ignition by the electric spark, and lasts up to the moment when the flame reaches the pipe walls. Details of the process can be seen from the slit scan in Fig. 9.2, from the frames *a*, *b* in Fig. 8.3 and frames 1–3 in Fig. 9.4. The frames in the figures have been obtained at different adjustment of the visualization system, so that there is some difference in their appearance. Anyhow it does not prevent us from realizing the essence of the modifications.

This stage is characterized by the flame origination, its front surface formation, and short period of the flame uniform propagation followed soon by strong acceleration and subsequent speedy small deceleration.

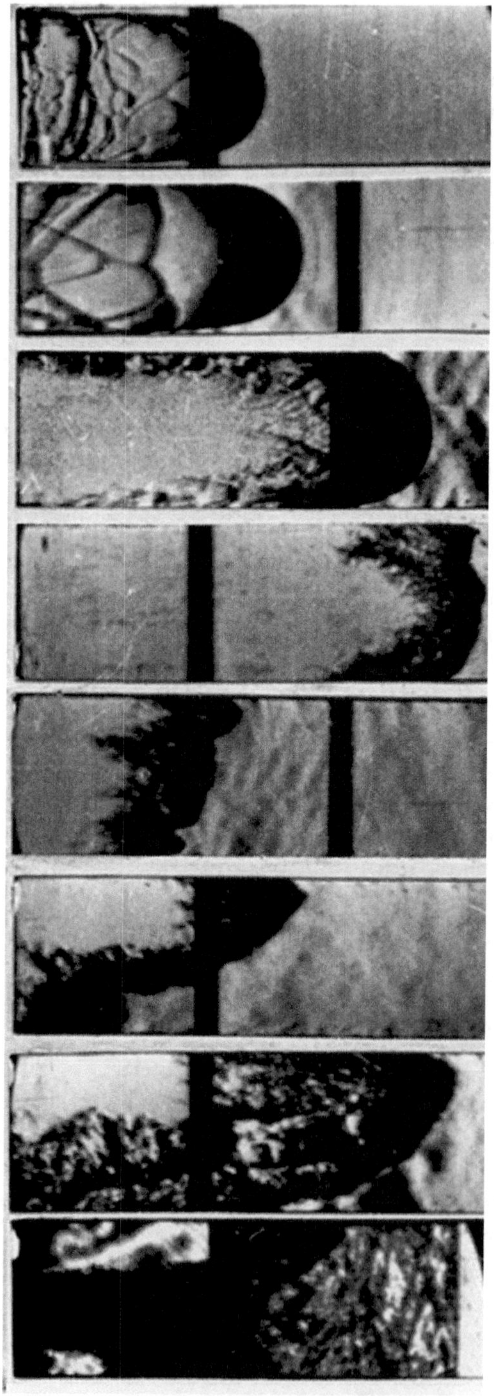

1.
X = 4.0 cm,
$\tau$ = 0.12 ms

2.
X = 15 cm,
$\tau$ = 1.00 ms

3.
X = 17.5 cm,
$\tau$ = 1.13 ms

4.
X = 26.5 cm,
$\tau$ = 1.54 ms

5.
X = 31.3 cm,
$\tau$ = 1.94 ms

6.
X = 41.1 cm,
$\tau$ = 2.16 ms

7.
X = 61.5 cm,
$\tau$ = 2.75 ms

8.
X = 113.7 cm,
$\tau$ = 3.43 ms

**Fig. 9.4.** Structures during the flame acceleration

The flame origin serves as the source of the weak pressure wave packet. The wave, being repeatedly reflected from the pipe walls, makes considerable impact on the flame front surface shape, and on its structure as well. The wave influence becomes apparent after some period, about 0.07 ms, during which the wave runs up to the walls and returns to the start point.

The flame holds a hemispherical configuration till the moment of the meeting with the wave reflected from the walls, and propagates in the pipe in the same way as in the free space until the meeting. During this phase the flame velocity $U$ and the velocity $W$ of the fresh gas convective motion are easy to express through the flame normal velocity $u_0$:

$$\left.\begin{aligned} U &= u_0\, n; \\ W &= U - u_0 = (n-1)\, u_0, \end{aligned}\right\} \tag{9.2}$$

where $n$ is the ratio of densities of the fresh mixture and of the combustion products. While the flame velocity is small in comparison with the sound velocity, as it is in this case, the density ratio can be replaced by the ratio of the temperatures T and molecular weights $\mu$:

$$n = \frac{\varrho_0}{\varrho_T} = \frac{T\,\mu_0}{T_0\,\mu_T}. \tag{9.3}$$

The thermodynamic calculation [53], for the chosen mixture initial temperature $T_0 = 298$ K ($\mu_0 = 29.85$) at the atmospheric pressure, gives the combustion products temperature $T = 2{,}777$ K and molecular weight $\mu_T = 26.7$, whence $n = 10.05$. Accordingly, the velocity of the flame propagation and the gas convective velocity make up $U \approx 23$ m/s and $W \approx 21$ m/s.

The flame meeting with the waves reflected from the walls causes redistribution of the gas convective velocity $W$, and there appears a preferred orientation of the gas motion along the pipe axis. Generally, the velocity of the flame and the gas in front of the flame are defined by formulas (9.2).

However when the flame surface area $S$ exceeds the area of the pipe cross-section $F$, the gas's effective convective velocity $W$ depends on the ratio of the areas $S/F$ and the direction. Averaging the second expression in (9.2) at constant pressure gives the velocity $W$ mean value in the $x$-axis direction

$$\overline{W} = \frac{S}{F}\,(n-1)\, u_0. \tag{9.4}$$

The product $S(n-1)u_0$ is merely growth of volume of the mixture combustion products due to combustion at the flame's enlarged surface area S at nearly constant pressure. At the beginning of the flame propagation, while the area $S$ is much less as compared with $F$, the flame front shape is deformed slightly and its linear dimensions vary proportionally to time $t$. Hence, $\overline{W} \sim t^2$ and the flame acceleration looks as $d\overline{W}/dt \sim t$, so that the pressure waves are generated along with the mixture ignition.

A question arises at this point as to whether a source of pressure waves is expansion of the combustion products, or energy release due to the electric discharge. The inquiry answer definitely follows from comparison of two sets of the frames presented in Fig. 9.5. The electric spark is seen to produce a pressure wave, though it is very weak and dies away, i.e., oversteps the limits of the equipment sensitivity, practically during 10 μs, at least. As to the wave produced by combustion, it is supported by expansion of the combustion products. In other words, the electric discharge does not influence noticeably the process development.

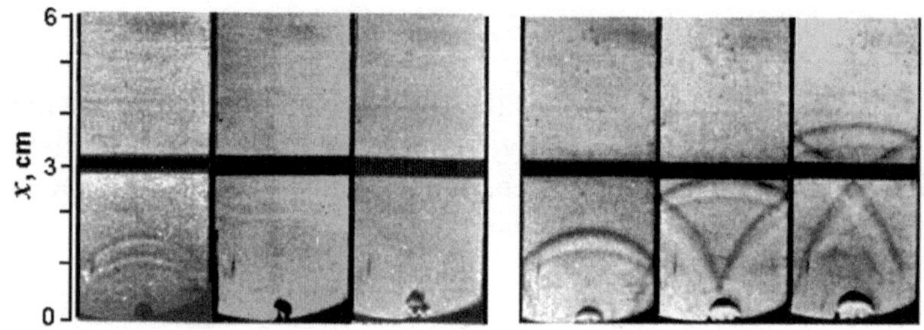


**Fig. 9.5.** Perturbations created by an electric spark in air (*left*) and an explosive mixture ignition (*right*): $n = 31.4 \times 10^3$ fr/s, $\tau = 4$ μs

Some time later, say, in 0.25 ms, the flame surface deformation stimulating expansion of the combustion products in the $x$-axis direction leads to a velocity increase up to 120 m/s. At the time, the flame front moves in the radial direction at a velocity of about 7 m/s. When the flame stretches along the pipe axis, the surface takes the form of a so-called cellular structure. The fact that the cellular structure appears already under the condition $S \ll F$ has been mentioned in Chapter 8.

The cellular structure exhibited by flames was especially studied by many authors (e.g. [54, 55, 73]), who gave different reasons for the origin of such a structure. As to the present research, the effect of wave perturbations of an acoustic nature is shown to be the real reason of the flame cellular structure formation. It distinctly follows from the Schlieren photos *a* and *b* in Fig. 8.3.

It is considered to be the case that the cellular structure does not influence the flame's normal velocity. Meanwhile, the mentioned photos do not confirm this supposition. Indeed, the flame should propagate in the direction of the normal to the envelope surface at the velocity of 23 m/s, whereas according to the graphs in Fig. 8.3, the value estimation for the flame at the cellular structure gives 34–35 m/s. This difference, though small, arises evidently from the growth of the total combustion rate because of the effective area increase in the case of the cellular flame surface.

A little more later, in $t = 1.2$ ms, the flame contacts the walls. After that the flame surface quickly begins to lessen, and the flame propagation along the pipe axis decelerates. Really, according to the photos *b* in Fig. 8.3, the angle between the tangent to the flame surface and the pipe wall surface can be estimated as $\phi \approx 30'$. If the flame moves to the wall at the normal velocity $u_0$, then the point of contact runs along the wall surface at a velocity which is equal to $U_W = u_0/\sin\phi \approx 250$ m/s. This results in a quick reduction in the flame surface.

That is way the flame front loses its velocity in such an up-tempo, just after the instant when the flame comes into first contact with the wall. The surface lessening lasts until the velocity $U_W$ is equal to the flame propagation velocity $U$,

$$U_W = \frac{u_0}{\sin\phi} = U. \tag{9.5}$$

The flame surface area $S$ reduction is the main reason of deceleration of the flow in front of the flame, and consequently, of the flow's mean velocity $\overline{W}$ decrease. Other reasons which are discussed in the literature (see, e.g. [66–68]) are most likely not so essential.

Some conclusions follow from formula (9.5). The angle $\phi$ which can be easily measured by an instant picture of the flame surface area provides us with information concerning the ratio of the normal and propagation velocities of the flame. However, note that the relation is valid while the pressure augmentation and flow velocity $\overline{W}$ are not too large, otherwise some other velocity values should be used in (9.2) and (9.4).

Meantime, there develop some other processes, so that the flame obtains a new state in which the shape, propagation velocity and gas mean velocity before the flame remain practically permanent. To the point, under the circumstances the flame operates as the piston mentioned in Chapter 4 that moves at the constant velocity $\overline{W}$.

## 9.4 Uniform flame propagation and second acceleration

The graph in Fig. 9.3 depicting the flame propagation velocity $U$ as a function of the distance $X$ can be attributed to the same dependence of Reynolds number $Re_d$ of the fluid flow before the flame front as well. The new ordinate transformation can be performed by means of expressions (9.2)–(9.4). According to these formulas, the flow velocity averaged over the pipe cross-section rapidly increases along with the ratio $S/F$.

When the ratio becomes equal to $S/F = 0.1$, i.e., shortly after the mixture ignition, the Reynolds number $Re_d$ reaches the critical value. Further, during the flame's uniform propagation, the Reynolds number reaches a quantity of the order $\mathrm{Re}_d \sim (3\text{–}4) \times 10^5$, which is 15–20 times as much as the critical value. On this ground the flame is presumed to move at this stage in the turbulent flow. And as the flame structure follows the structure of the flow, it is usually accepted as a turbulent flame ([49, 50], etc.). The same standpoint concerning the problem of flame acceleration, while the flow fluid propagation remains approximately uniform, is applied in the work [74] as well.

In reality, it does not seem to be the case. Let us look at the picture of the slit scan of the flame at the stage of uniform propagation, $x \in [18, 37]$ cm, in the square cross-section pipe shown in Fig. 8.2. Incidentally, waves reflected from the glass-wall joints are not seen, as stated above.

The gas mixture before the flame does not move along all of the length of the pipe, but only in the limits of the dilative complex named above as the double saltus. The gas mixture at some cross-section of the pipe begins moving when the first wave comes to the cross-section. The velocity is constant across the flow at the motion start, however then a new velocity profile is settled according to the flow's Reynolds number in the course of time.

If the time is sufficiently long, the flow and the flame as well become turbulent. Otherwise both of them remain laminar. However, there can happen an intermediate phase, when the perturbation waves exist in the flow, but the dynamic equilibrium between the perturbation birth and damping, and the velocity profile, have no time to come to the settled state.

The length of the turbulence coming to steady state depends on the flow mode [8]. In the laminar flow this process occurs at $x \approx 0.04 d\mathrm{Re}_d$. The length of the turbulent flow formation is different depending on, probably, the feature of the experiment specifics. For instance, J. Nikuradse [40] estimates it as 40 calibers, while according to G. Kirsten's measurements [75], the length makes up 50–100 calibers.

The length in our case, with the pipe cross-section area equivalent to 3 cm diameter, should exceed 120 cm, at least, whereas the length of the double saltus, estimated by the slit scans shown in Figs. 9.2 and 8.2, does not exceed 30 cm. Consequently we deal just with the intermediate case, when the flow velocity profile and oscillation radial distribution have been formed. In the course of time the double saltus length increases, so the flow velocity and oscillation profile evolve. The photos 1–5 in Fig. 9.4 as well as frames *b*, *c* in Fig. 8.3 illustrate the evolution process.

In the beginning we see that there appear gradually a considerable quantity of disturbance waves in front of the flame, though the boundary layer is not seen as yet. Meanwhile photos 4, 5 (Fig. 9.4) and frames *c* (Fig. 8.3) reveal some essential changes in the flame shape and structure. The enveloping surface $S$ of the smooth flame of prolate configuration becomes equal only to $S \approx 2.5F$, and the question is how the flame propagation is supported at nearly constant velocity. In addition, what is the reason of the flame's second acceleration.

Change in the flame structure provides an answer to the question. As the photos in Fig. 9.4 demonstrate, the flame length shortening is accompanied by change in the inner structure which is converted into the one looking like a turbulent structure. Effective combustion surface $S_{\mathrm{eff}}$ of the turbulent flame is known to exceed noticeably the flame smooth surface $S$.

Thus, the surface enveloping the flame contour is strongly decreased. The area size measurements, say, in the photos 2 and 3 (Fig. 9.4) show, that the flame enveloping surface area decreases from approximately $S_2/F \approx 30$ to about $S_3/F \approx 2.5$. Nevertheless, the flame propagation velocity deceleration amounts only to about 20%. Hence the decrease of the flame surface area is countervailed owing to the structure change that gives nearly tenfold increase in the combustion's effective surface, which is $S_{\mathrm{eff}}/F \approx 10$. This is the reason for the flame propagation at nearly the same velocity.

The flame of the changed structure might well propagate at permanent velocity. However the structure should be steadily maintained by the pressure waves. We have seen from oscillograms in Fig. 8.1 and know from the paper [73], that energy output in the zone of burning is the source of pressure waves. Since the double saltus lengthens in the course of flame propagation, the combustion zone becomes a more effective source of pressure waves. Therefore, the ratio $S_{\mathrm{eff}}/F$ and combustion intensification provide further augmentation. This process is illustrated with the frame series of high-speed filming shown in Fig. 9.6.

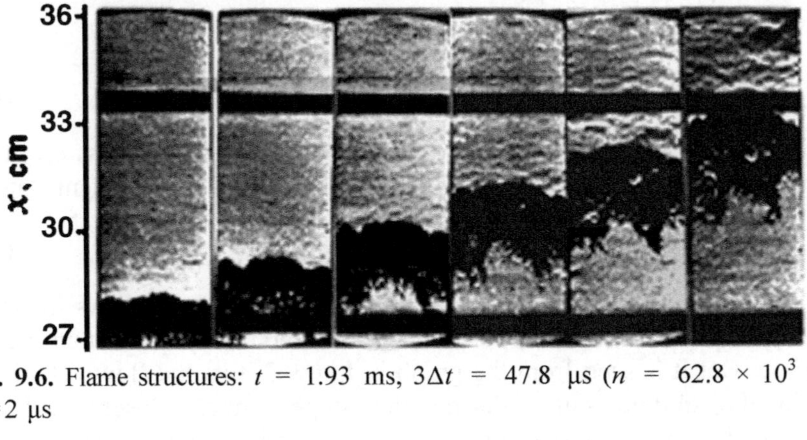


**Fig. 9.6.** Flame structures: $t$ = 1.93 ms, $3\Delta t$ = 47.8 μs ($n$ = $62.8 \times 10^3$ f/s), $\tau$ = 2 μs

The frames in the series have been made with an exposure duration of $\tau$ = 2 μs, and a frame-repetition period $T$ = 15.9 μs at a picture frequency $n$ = $62.8 \times 10^3$ f/s (frames per second). Only every third frame has been included in the series, so the time interval between the frames shown corresponds to $3\Delta t = 47.8$ μs. The flame appears in the window's field of vision in $t \sim 1.8$ ms after the mixture ignition. Naturally, all frames of the series have been registered with the equipment at equal sensitivity adjustment. This identity allows the process development to be considered in detail.

Comparison of the scenes on the frames clearly shows intensification of the perturbation waves in front of the flame, as well as arising of the prominences on the flame surface near the walls. These events result from the aforesaid double saltus progressive lengthening. It was noted in [76], that the double saltus length is to some extent similar to the length of a plate blown by a stream. As the flow length increases within the double saltus, there appears in the flow a laminar boundary layer which comes into the turbulent layer in the course of the double saltus lengthening.

The parameters of the boundary layer in the gas flow behind a shock wave were estimated in the work [77] according to results of measurements in a flow over a plate. By analogy, we use the data from [8] on the subject for the estimation of the boundary layer parameters in the double saltus, provided that the plate length is equal to the length $l$ of the double saltus.

The laminar boundary layer at the plate surface grows into the turbulent layer, when the flow Reynolds number composed for the plate length $l$ attains the value of the order $\mathrm{Re}_l = \mathrm{Re}_{\mathrm{cr}} = Wl/\nu \sim 10^6$. Whence the critical length $l_{\mathrm{cr}}$ for the laminar boundary layer can be derived in the form

$$l_{\mathrm{cr}} \approx 10^6(\nu/W). \tag{9.6}$$

On the other hand, the thickness of the laminar and turbulent boundary layers are approximated by the relations:

$$\delta_{\mathrm{lam}} \approx 5\left(\nu/W\right)\sqrt{\mathrm{Re}_l}\,, \tag{9.7}$$

$$\delta_t \approx 0.37\, l_t \left(\mathrm{Re}_l\right)^{-1/5}. \tag{9.8}$$

It gives an opportunity to estimate the boundary layer thickness $\delta_{\mathrm{cr}}$ at the length $l_{\mathrm{cr}}$ when the laminar layer is about to grow into the turbulent state:

$$\delta_{\mathrm{cr}} \approx 5\left(\nu/W\right)\sqrt{\mathrm{Re}_{\mathrm{cr}}}\,. \tag{9.9}$$

Generally speaking, the cited formulas concern those flows in which velocity is permanent. In the case under consideration this is not absolutely the case. But it is seen from the graph in Fig. 9.3, that the flow velocity $W$ within the double saltus during the stage agrees closely with the requirement. Taking into account these cases and relations (9.2), (9.4), the effective velocity can be estimated as $W = 120$ m/s. The kinematic viscosity factor $\nu$ for the chosen mixture is known, $\nu = 0.15$ cm$^2$/s. Then the length $l_{\mathrm{cr}}$ and thickness $\delta_{\mathrm{cr}}$ of the laminar boundary layer near the place of conversion into the turbulent state can be estimated by formulas

$$l_{cr} \approx \frac{\nu}{W} \mathrm{Re}_{cr} \sim 13cm, \tag{9.10}$$

$$\delta_{cr} \approx 5 \frac{\nu}{W} \sqrt{\mathrm{Re}_{cr}} \sim 0.7mm. \tag{9.11}$$

These data help to determine the place of the conversion in the process under consideration. Looking at the slit scans in Figs. 9.2 and 8.2, we see that the double saltus has the length $l_{cr}$ = 13 cm. The length corresponds to the flame's front position at the distance $x$ = 22 cm from the ignition point. Consequently, this is that flame position from which the turbulent boundary layer arises.

Now, as it has been shown above (see Fig. 8.12), the outside surface of the boundary layer is characterized by a sharp increase of the flow parameter oscillations. It is of interest to estimate an effect of the oscillations on the flame velocity, which is known to increase noticeably in the presence of the oscillatory motion.

Among a great number of formulas offered for the turbulent flame velocity $u_t$ (see, e.g. [68, 78–80]), we will use the most simple expression that is sufficient for our purpose: $u_t = u_0 + w$. The oscillation amplitude is denoted here as $w$. Then the propagation velocity of the turbulent flame $U_t$ takes the form

$$U_t = W + w + u_0. \tag{9.12}$$

If the turbulent boundary layer arises immediately in front of the combustion zone, then the amplitude $w$ of the velocity oscillations and the flame velocity $U_t$ have maximum values at the distance $\delta_{cr}$ from the wall. That means according to (9.12), that prominences appear on the flame surface at the distance $\delta_{cr}$ from the walls. The beginning of such a prominence is demonstrated in frame 4 in Fig. 9.5.

Following [74], the local maximum of the flame velocity we will call "a leading point". The point moves faster than remaining parts of the flame surface and determines the propagation velocity profile. But while the length of the turbulent part of the boundary layer is small still, it burns out fast, which results in emission of a compression wave. Since no modifications happen to the flame, total velocity remains practically constant on average. However in actuality the flame velocity makes some small oscillations about the average value.

The process of a leading point formation is illustrated by the outline in Fig. 9.7. As is seen on the picture, the same leading point appears at the other wall of the pipe. The outline represents the flame propagation process in one plane. Meanwhile, it is easy to fancy, that such leading points originate along all the periphery of the pipe cross-section. Therefore we can think about the leading ring moving close to the walls, which the photos in Fig. 9.6 confirm.

However the uniform ring distribution is unstable and can not exist for a long time. One of the points moves forward, as the sketch in Fig. 9.7 illustrates, and the frames shown in Fig. 9.8 verify. The point's subsequent progress leads to the flame asymmetry. This process is demonstrated by the frames in Fig. 9.9.

Henceforward the point influence on the flame increases progressively. Accordingly, the flame is taking an asymmetrical shape, and its surface is growing in area. This process development can be traced by the frame series in Figs. 9.7 and 9.9. Similar flame configurations at the non-stationary propagation were observed in a number of researches (see, e.g. [81–83]).

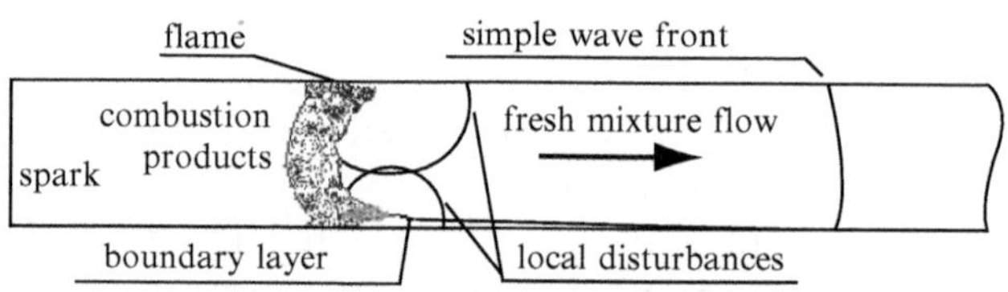


**Fig. 9.7.** Diagram of quasi-stable double saltus

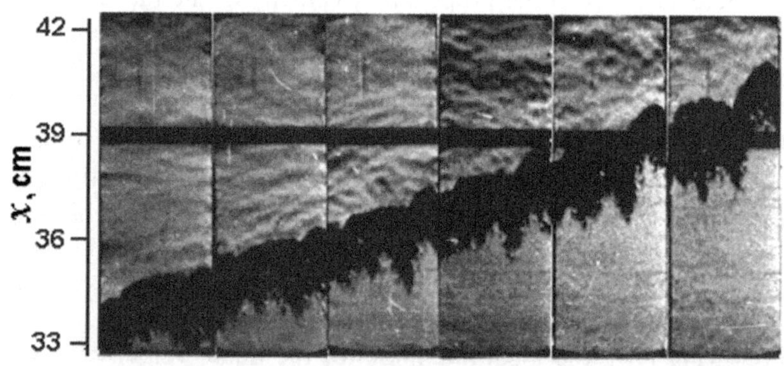


**Fig. 9.8.** Flame structures: $t = 1.93$ ms, $3\Delta t = 47.8$ μs ($n = 62.8 \times 10^3$ f/s), $\tau = 2$ μs

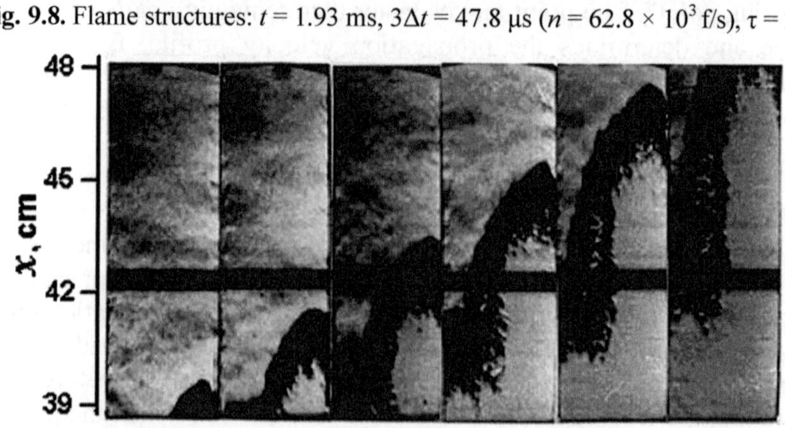


**Fig. 9.9.** Flame structures: $t = 2.14$ ms, $3\Delta t = 63.7$ μs ($n = 47.1 \times 10^3$ f/s), $\tau = 3$ μs

An interesting visual demonstration of the leading point growth in development of a turbulent boundary layer was described in [78]. The experiments were performed for a water flow in an open rectangular channel $d = 5$ cm wide. There were measured the flow velocity averaged in time, $\overline{U}$, and the root-mean-square amplitude of the velocity oscillations, $\sqrt{\overline{u^2}}$. In the course of the turbulence development while the water moved down the flow, the maximum of the sum $S = (\overline{U} + \sqrt{\overline{u^2}})/U_0$ shifted along with the thickness of the boundary layer. It is the main point of the above reasoning.

Profiles of the relative velocity $\overline{U}/U_0$ and sum $S$ are represented by the graph in Fig. 9.10 as the dependences on the distance Δ from the middle of the channel for the flow's two cross-sections characterizing the process of closing of the opposite boundary layers. The curves on the left relate to the initial closing stage, when the boundary layer thickness is $\delta = d/6$. The curves on the right image the process at a later stage, when the boundary layer edge is found at the distance $\delta = d/3$ from the wall.

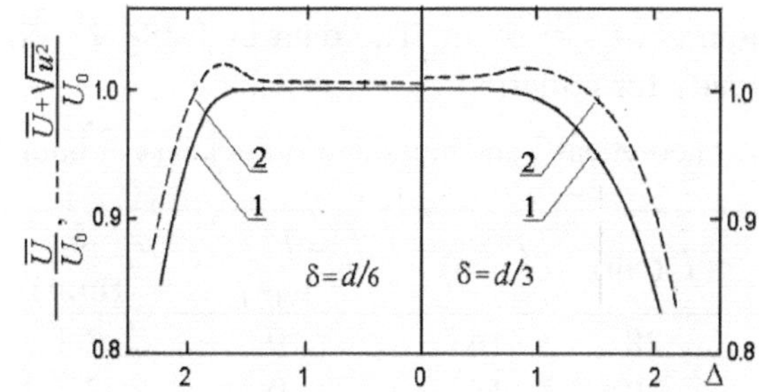


**Fig. 9.10.** Profiles of velocity $U/U_0$ (1) and sum $S = (\overline{U} + \sqrt{\overline{u^2}})/U_0$ (2) for two cross-sections of water flow in an open channel [78]

Data of these experiments confirm the notion according to which the maximum of velocity sum $S$ is localized at the edge of the boundary layer. The quantity $S$ is nothing but the sum of two decisive terms in formula (9.12) for the flame velocity. Hence, we may state on the basis of the data and the results [84–86], that the most velocity of the flame propagation is situated near the wall, just at the edge of the boundary layer where the leading point emerges. The leading point determines the flame shape and the inner structure that are responsible for the flame's second acceleration.

When the second acceleration comes into force, the pressure waves emitted by the flame during the first acceleration are far ahead, in the region of the double saltus front. The accelerating flame at this stage creates within the double saltus some new system of pressure waves of stronger intensity caused by progressing intensification of the mixture combustion.

These waves interact with the boundary layer in different ways, in particular, not excepting direct separation of the layer. Therefore, simple application of relations (9.10), (9.11) in such conditions seems to be completely unfit for calculation of the length $l_{cr}$ and critical value $\delta_{cr}$ of the laminar boundary layer (see Chapter 8).

In spite of the fact that the calculation of the length $l_{cr}$ turns into a rather complicated problem for the double saltus of the actual configuration, some opportunity for an estimation exists. The problem of the estimation is reduced to the length $l_{cr}$ variation and finding such a length value at which the boundary layer thickness $\delta$ computed according to equation (9.8), and the flame's leading point distance $\Delta$ from the wall, measured in experiment, would have the closest values.

Photos presented in Figs. 9.5, 9.7, 9.8 and 9.9 have been used in measurements of the parameters $L$, $l$ and $\Delta$. On the other hand, the calculations have been performed by the method of sequential approximations with the length $l_{cr}$ as the variable argument. The values of $L$, $l$, necessary for the calculations, have been taken from measurement of appropriate photos. The best coincidence has been found for the value of the variable argument $l_{cr}$ = 5 cm. The data in Table 9.1 correspond to the value, and are ready for collation.

**Table 9.1.** Theoretical $\delta$ and measured $\Delta$ thickness of boundary layer

| Fig. # | $L$ (cm) | $l$ (cm) | $l_t = l - l_{cr}$ (cm) | $\delta$ (mm) | $\Delta$ (mm) |
|---|---|---|---|---|---|
| 9.5 | 28 | 14 | 9 | 2.1 | 2.0 |
| 9.7 | 28 | 14 | 9 | 2.1 | 2.0 |
| 9.5 | 34 | 18 | 13 | 2.8 | 2.5 |
| 9.7 | 34 | 18 | 13 | 2.8 | 2.5 |
| 9.8 | 36 | 20 | 15 | 3.2 | 3.1 |
| 9.8 | 45 | 25 | 20 | 3.9 | 4.0 |
| 9.10 | 55 | 29 | 24 | 4.7 | 4.8 |

Coincidence of the boundary layer thickness calculated for $l_{cr}$ = 5 cm, and the leading point shift observed in experiment means, that it is turbulization of the flow in the boundary layer of the double saltus that may be responsible for the flame's second acceleration. Though it might be not the only cause that contributes to the flame acceleration development. The slit scan in Fig. 9.11 and series of frames in Fig. 9.12 help to comprehend the additional factor promoting the acceleration development.

The sharp increase in flame-piston velocity and reinforcement of the flow disturbances lead to some curious results. The gas mixture is not consumed completely within the flame front. It continues burning far behind the front. The flame itself breaks down into small cells and turns into many subsections within a lengthy zone of combustion, as seen in the photos in Figs. 9.11 and 9.12. The length might run up to 20 cm.

**Fig. 9.11.** Slit scan of double saltus during evolution of a second flame acceleration

Such a strong change of the flame structure results in multiple increase of total rate of the chemical reactions and the energy release, that call into play the flame's further front acceleration. Note, the enveloping flame surface is still formed according to the concept of a leading point in these new conditions, as the frames in Fig. 9.12 show.

**Fig. 9.12.** Flame structures: $t = 2.72$ ms, $3\Delta t = 47.8$ μs ($n = 62.8 \times 10^3$ f/s), $\tau = 2$ μs

## 9.5 Formation of detonation wave

We have just seen that the flame configuration, shape and structure, has a decisive effect on the double saltus properties which depend on flame acceleration. The quantitative dependence of the flame velocity changes on the distance covered can be deduced from function $U(X)$ plotted in Fig. 9.3.

Differentiating this function with respect to $X$, we will obtain dependence $dU(X)/dX$ plotted in Fig. 9.13, equivalent to the flame acceleration.

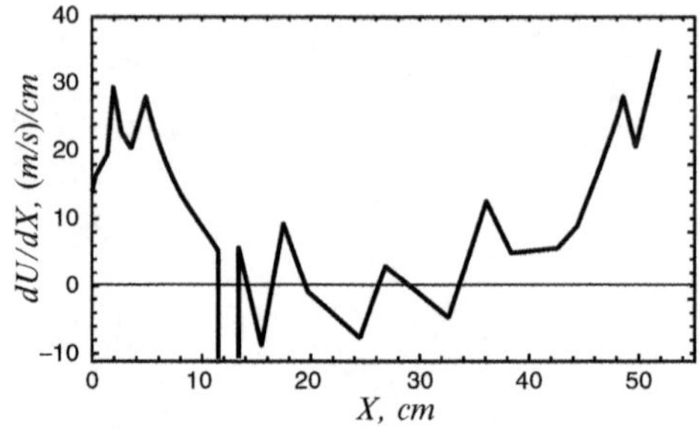


**Fig. 9.13.** Flame velocity change as function of the path covered

On the graph are well distinguished the regions of the first acceleration, the next short strong deceleration, motion at approximately constant velocity, and the second acceleration. Both accelerations appear to have considerable values. However the first acceleration is caused by the motion of the flame front along the slot for observation. This occurs due to transition from the volume expansion of the combustion products to the lengthening along the slit. The main reason of the second acceleration is attributed to the combustion's total rate increase, when the flame's effective surface rises steeply under the action of the augmented compression waves.

The first acceleration creates the double saltus, from which the forward front moves much faster then the flame. When the second acceleration starts, the forward front leaves the flame ~46 cm behind. The pressure in the double saltus at the time is found to be at the level of 2 atm, as it follows from the oscillograms in Fig. 8.1.

During the second acceleration, especially at a later stage, intensive compression waves emitted by the flame tend to merge rapidly, forming a powerful shock wave moving about 6 cm before the flame front. In this way there appears some new double saltus within the previous one. The pressure within the new double saltus reaches 10–11 atm. Formation and development of the new structure are demonstrated by the frame series in Figs. 9.14 and 9.15.

These photos give evidence that the distance length between the new shock wave and the flame becomes gradually shorter, and it is curtailed from 6 to 2 cm. Hence, the flame acceleration prolongs its increase at an even faster rate. As the shape of the flame's surface and volume of the combustion zone are practically not changed, the acceleration augmentation may be explained as nothing but total burning velocity in the flame volume. Thus temperature of the mixture rise behind the new shock wave causes the acceleration at this stage of the double saltus development.

For estimation of these newly arisen requirements, we will take advantage of the data given in [87] concerning dependences of the shock wave velocity $D$ and velocity $W$ of the gas behind the wave on the pressure jump $P_1/P_0$ at the wave. The parameters calculated in [87] for air with initial standard conditions ($P_0 = 1$ atm, $T_0 = 273$ K) are presented in Table 9.2.

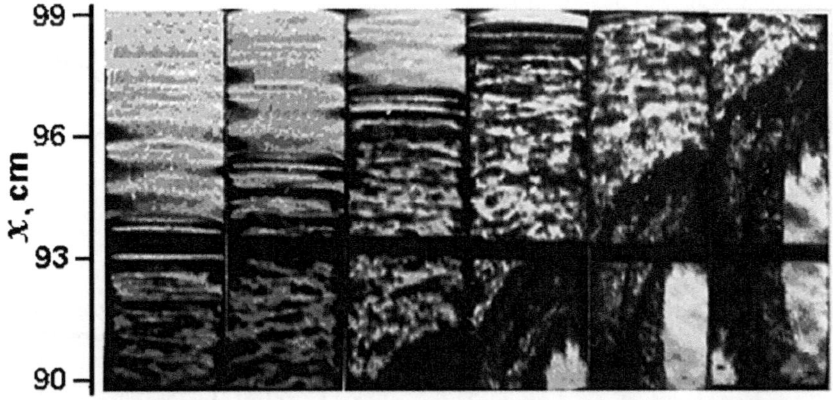


**Fig. 9.14.** Flame structures: $t = 3.26$ ms, $3\Delta t = 32$ μs ($n = 125.7 \times 10^3$ f/s), $\tau = 1$ μs

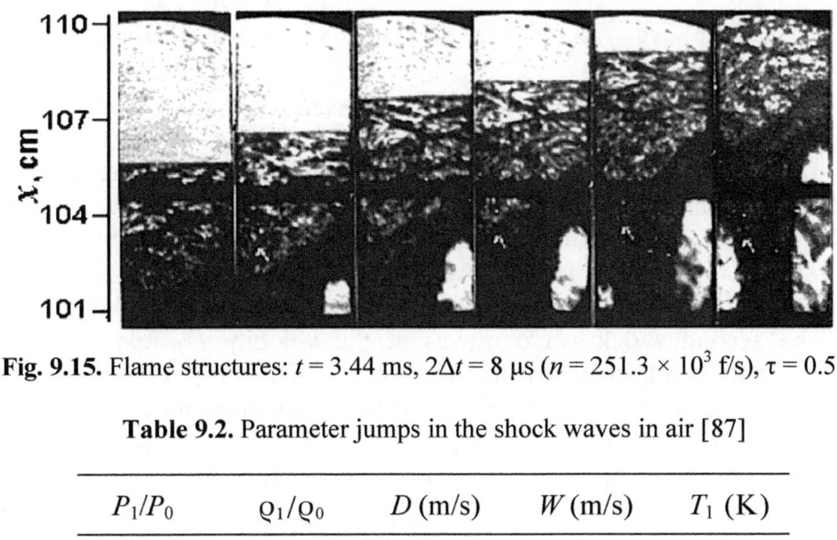


**Fig. 9.15.** Flame structures: $t = 3.44$ ms, $2\Delta t = 8$ μs ($n = 251.3 \times 10^3$ f/s), $\tau = 0.5$ μs

**Table 9.2.** Parameter jumps in the shock waves in air [87]

| $P_1/P_0$ | $\varrho_1/\varrho_0$ | $D$ (m/s) | $W$ (m/s) | $T_1$ (K) |
|---|---|---|---|---|
| 2 | 1.63 | 452 | 175 | 336 |
| 5 | 2.84 | 698 | 452 | 485 |
| 10 | 3.88 | 978 | 725 | 705 |
| 15 | 4.35 | 1,196 | 943 | 940 |
| 25 | 5.00 | 1,545 | 1,270 | 1,350 |
| 50 | 6.04 | 2,150 | 1,795 | 2,260 |

The table data seem to be quite applicable to estimate the main features of the process under consideration for further development. Indeed, initial

differs for both gases by not more than 3%. Though data in the table have been calculated with rather large steps and need interpolation. The interpolated dependences are represented by appropriate graphs in Fig. 9.16.

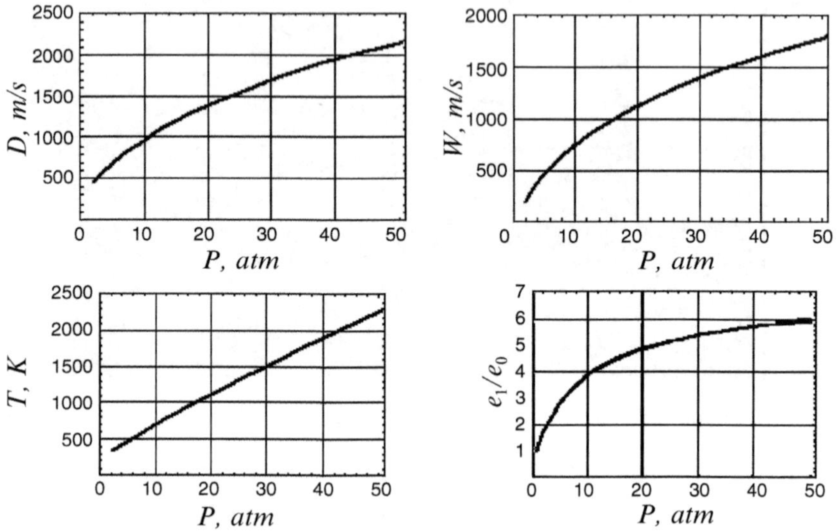


**Fig. 9.16.** Dependence of shock wave parameters on the pressure jump

The gas mixture flow within the double saltus may be represented by a somewhat simplified model as follows. The first shock wave, i.e., the forward front of the double saltus, propagates in an immobile gas at the velocity $D_1$. The gas in the flow created by the wave and supported by the flame is heated up to the temperature $T_1$ and moves at the velocity $W_1$.

The second shock wave moves at the velocity $D_2$ relative to the medium, and its total velocity amounts to $D = D_1 + W_1$. The velocity $W$, pressure $P$ and temperature $T$ of the gas being not consumed still increase according to $W = W_1 + W_2$, $P = P_1 + P_2$, $T = T_1 + T_2$. The flame's normal velocity $u_0$ in the compressed and heated gas mixture should grow, for the chemical reactions to be brought to a point any quicker. The growth of chemical reaction rate occasioned by the gas mixture compression and heating should cause a rise in the flame's normal velocity $u_0$ and, certainly, the propagation velocity $U$ of the flame front which more precisely to define now as the combustion zone front.

The flame's normal velocity $u_0$ depends on many factors. Their effects were considered in detail in the literature (see, e.g. [66–68, 88, 89]). When the mixture consists of some fixed components, as is the case in question, the dependence can be written in rather simple form:

$$u_0 \sim \frac{T}{\sqrt{P}} \sqrt{\frac{g(T_{si})}{T_{si}\left(T_{si} - T\right)}} . \tag{9.13}$$

Here $T_{si}$ is the temperature of spontaneous ignition of an explosive mixture, $g(T_{si})$ denotes an effective rate of the combustion's chemical reactions.

The formula allows us to interpret the concept of the spontaneous ignition temperature as follows. When the mixture temperature comes to the spontaneous ignition temperature, $T \to T_{si}$, the flame's normal velocity $u_0$ grows exponentially, theoretically without limit, and the chemical reaction actually develop simultaneously throughout all the volume of the heated mixture.

A distinctive feature of this process is a harshly accelerating transition from slow development of the chemical reaction, as the first phase, to its sharply explosive course of the energy release, as the second phase. The time interval in which the chemical reaction does not evince any perceptible calorific effect is usually termed as the period of induction delay $\tau_i$. The curves on the graph in Fig. 9.17 illustrate the physical meaning of the induction delay.

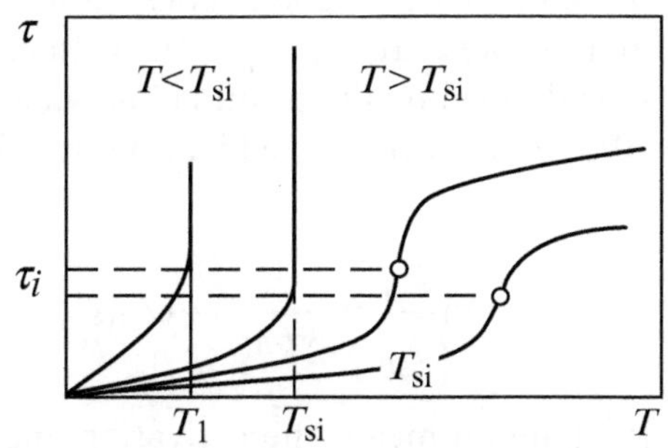


**Fig. 9.17.** Dependence of induction delay on temperature of explosive mixture during chemical reactions

While the mixture temperature is relatively low, $T < T_{si}$, the time delay in chemical reaction progress is too long to be observed, and the temperature practically does not exceed some value $T_1$. The spontaneous combustion and temperature increase, i.e., realization of the second phase, does happen under the condition of $T > T_{si}$.

The time delay $\tau_i$ dependence on the mixture temperature $T$ at this stage is approximately identical to most parts of explosive gases [65]. This dependence, mentioned in many papers (see, e.g. [49, 90, 91]) is expressed by the formula:

$$\tau_i \sim \exp\left(E/RT\right). \tag{9.14}$$

One of the versions of the function $\tau_i(T)$ checked in experiments with the mixture of methane with oxygen $3CH_4 + 2O_2$ at atmospheric pressure [67], closely related to the case similar to the one under consideration, is reproduced in Fig. 9.18.

The formula (9.14) allows us to find an interdependence between the spontaneous ignition temperature $T_{si}$ and induction period duration $\tau_i$. For example, the induction period $\tau_i = 100$ μs implies the spontaneous ignition temperature being equal to $T_{si} = 1{,}550$ K.

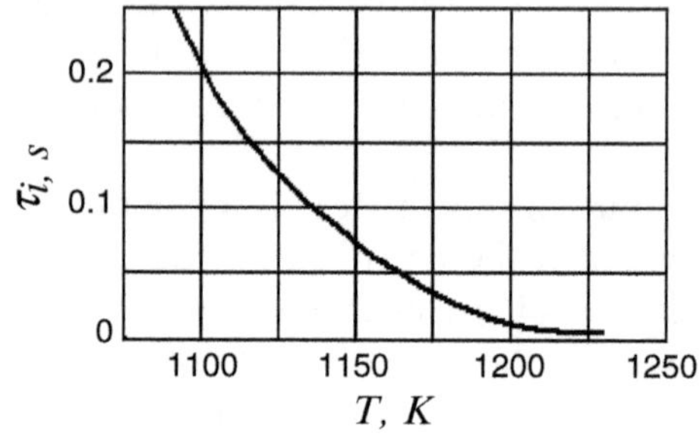


**Fig. 9.18.** Experimental dependence $\tau_i(T)$ for mixture $3CH_4 + 2O_2$ [67]

Here we should note that the spontaneous ignition temperature $T_{si}$ is a function of the mixture pressure $P_{si}$ ([92, 93]). Interdependence between the two parameters is defined by the formula known as Semenov's relation. With the view of the research on the subject the relation can be written as follows

$$\ln \frac{P_{si}}{T_{si}^2} \approx \frac{E}{2RT_{si}} + \text{const}, \tag{9.15}$$

where $E \approx 5 \times 10^4$ kal/mol means the activation energy, and $R$ is the gas constant.

Rewriting (9.15) for $T_{si} = T + \Delta T$, where $\Delta T$ is caused by the pressure increase above $P_0$, we will obtain the equation in the form of the implicit function $\Delta T(P)$ which is plotted in Fig. 9.19:

$$P = \left(\frac{T - \Delta T}{T}\right)^2 \operatorname{Exp}\left(\frac{E}{2R} \frac{\Delta T}{T(T - \Delta T)}\right), \tag{9.16}$$

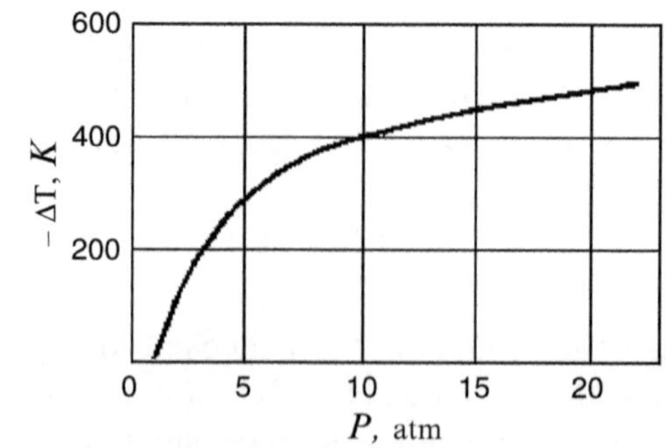


**Fig. 9.19.** Ignition temperature shift $\Delta T$ as function of pressure $P$

The data presented by the graphs in Figs. 9.16, 9.18, 9.19, and by formulas (9.14)–(9.16) give an opportunity to estimate the explosive mixture state within the region of greatest compression, i.e., within that inner double saltus. The state is defined by such parameters as the velocity $D_2$ of the second shock wave, total pressure $P$, temperature $T$, and by duration of the flow element being in the region until the flame consumes it. The residence time $\tau$ is evaluated according to the relation $\tau = l(D_2 – W_2)$.

As is seen from Table 9.2 and from graphs in Fig. 9.16, the first shock wave has velocity $D_1 \approx 450$ m/s and pressure jump $P_1/P_0 \approx 2$. The wave produces the flow of the mixture moving at velocity $W_1 \approx 175$ m/s. Similar data concerning the flow behind the second shock wave can be derived from photos shown in Figs. 9.12, 9.14 and 9.15.

The second shock wave moves at the moment of its formation at velocity $D \approx 1{,}000$ m/s which is incremented to $D \approx 1{,}300$ m/s in a short while. Despite such a quick augmentation of the velocity $D$, the length $l$ of the second double saltus decreases from 4.7 to 2 cm during that short time. Hence the flame is accelerated even more rapidly then the shock wave. As a matter of fact it is the third acceleration of the flame.

**Table 9.3.** Flow parameters in compression zone $l \in [4.7, 2]$ cm

| $D$ (m/s) | $W_2$ (m/s) | $T_2$ (K) | $P_1 + P_2$ (atm)_ | $T_1 + T_2$ (K) | $D_2 - W$ (m/s) | $\sim l$ (cm) | $\tau$ (μs) |
|---|---|---|---|---|---|---|---|
| 1,000 | 590 | 600 | 9.5 | 936 | 235 | 4.7 | 200 |
| 1,100 | 685 | 670 | 10.4 | 1,006 | 240 | 3.8 | 158 |
| 1,200 | 775 | 756 | 12.4 | 1,092 | 250 | 2.9 | 116 |
| 1,300 | 850 | 830 | 14.1 | 1,166 | 275 | 2.0 | 73 |

For ascertainment of the essential reason of the flame's third acceleration formation, we estimate the flow parameters in the compression zone for the velocity $D$ in the interval of $D \in [1{,}000, 1{,}300]$ m/s and the length $l$ in the limits $l \in [4.7, 2]$ cm. Table 9.3 demonstrates the estimation result. The data in the table testify to an appreciable increase in the mixture temperature at, pay attention, simultaneous abridgment of the mixture residence time $\tau$ within the second double saltus, i.e., in the flow of the most compressed mixture.

As has been estimated above, the induction time $\tau_i = 100$ μs is well matched with the spontaneous ignition temperature $T \approx 1{,}550$ K for the mixture at atmospheric pressure. Now, taking into account the induction time dependence on the pressure, the temperature correction for the pressure 14 atm by the formula (9.16) gives the temperature $T_{si} = 1{,}110$ K.

According to Table 9.3, the requirement $T > T_{si}$ is unconditionally fulfilled at the shock wave velocity $D \geq 1{,}200$ m/s. Then the distance $l$ between the second shock wave and the flame's leading point (the leading edge of the chemical reaction zone which the flame now is essentially turned into), is equivalent to the time induction delay value $\tau_i$. As a result, the shock wave and the combustion zone following behind the wave at a distance $l$ constitute a quasi-stationary complex united by some kind of feedback.

Indeed, the space within this complex is filled up by pressure waves emitted during the combustion process. The waves support the cycle including the second shock wave amplification and the combustion process intensification. The latter, in its turn, amplifies generation of the pressure waves. This interdependence was formerly noted by A.S. Sokolik [65].

The shock wave intensity and the double saltus parameter strengthening in the course of the complex relatively smooth development would reach the level at which the complex is converted into a detonation wave. However the detonation arises in some different way.

As it follows from the measurement results obtained by J. Campbell and considered in the books [67] and [68], the detonation wave in the mixture $CH_4 + 4O_2$ at $P_1 = 1$ atm, $T_1 = 293$ K is characterized by the parameters:

$$D_2 = 2{,}166 \text{ m/s}, P_2/P_1 = 26, T_2 = 3{,}860 \text{ K}. \qquad (9.17)$$

Comparison of these parameters with the experimental data for the shock waves from Table 9.3 shows, that the tabulated data are only about a half of the values shown in equation (9.17). Nevertheless, the double saltus develops into the detonation by some miracle, in spite of such a great difference in the parameter values. Anyhow, the frame series in Fig. 9.20 testifies to the double saltus transition into detonation. Such a "premature" formation of the detonation wave might be related to the second double saltus features determined by the peculiarity of the acoustic waves propagating in channels of finite aperture, considered in Chapter 4.

High temperature of the gas within the quasi-stationary complex reduces the picture contrast owing to the Schlieren-method specificity of visualization. Notwithstanding, there can be distinguished on the photo series in Fig. 9.20 the burning zone, its approach to the shock wave and the pressure waves emitted during combustion. The pressure waves are clearly seen to have appreciable curvature and to reflect from the pipe walls. Eventually, the mixture at a spot of one of the reflections is heated to such an extent (see Fig. 4.2), that thereat the chemical reaction creates a separate detonation wave. Two such detonation waves can be discerned at the left and right walls on frames in Fig. 9.20.

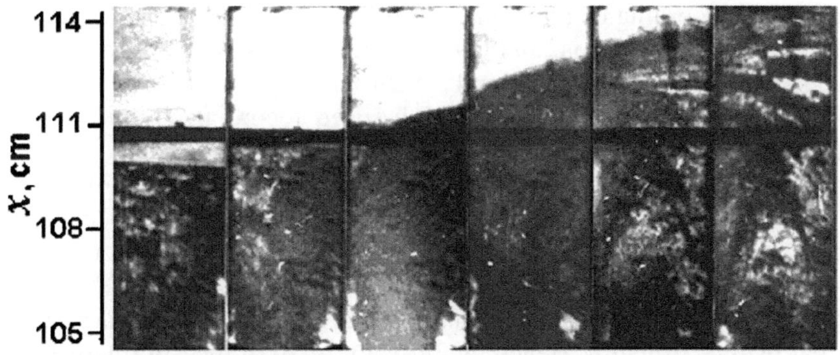


**Fig. 9.20.** Process of double saltus transition into detonation: $T = 3.45$ ms, $\Delta t = 4$ μs ($n = 251.3 \times 10^3$ f/s), $\tau = 0.5$ μs

The detonation, as nearly a point explosion, propagates from the wall in the form of a spherical wave. This wave propagating in the pipe is reflected from the walls just in the same way as it is shown in Fig. 6.2. The wave sector directed forward merges soon with the other detonation wave, and the wave pair becomes gradually more and more flat, moving on the fresh mixture as a usual detonation wave.

The opposite sectors of the waves move through the combustion products. Now and then it is called a "returnation" wave. The central sectors of the waves move through the environment mixture in which the chemical reaction continues. These parts propagate across the pipe, i.e., in a rather broad interval of directions around the perpendicular to the pipe axis, similar to the diagram in Fig. 5.2. Being reflected by turns from the walls, these parts of the ex-detonation wave leave legible intersecting traces on the frame photos.

Photo registration of the detonation wave at the moment of its formation is not a very simple operation. The problem consists in synchronization of the process and the exposition instant. The matter is that interaction of the acoustic waves with the flame during the flame evolution is a multivariant process which implies the dependence $X(t)$ being a stochastic one to some extent. At early phases of the development, the flame propagates slowly, and the probable deviations from the function can scarcely be observed. So that there is no problem with the synchronization, despite the limited frame number available in case of optical–mechanical unit application for high-speed recording.

Impotence of these deviations adversely affects the function stability, when the flame propagation velocity grows. This results in alteration of the detonation transition place and time, which vary within the ranges

$$X \in [105, 115] \text{ cm and } T \in [3{,}440, 3{,}450] \ \mu\text{s}. \tag{9.18}$$

Meanwhile recording of fine details of the process in the final development phase requires high spatial and temporal resolutions, and accordingly, high precision in large scale images and synchronization.

When an optical–mechanical high-speed recording unit is used, the synchronization is accomplished by the registration system, when the time reading starts from the mixture ignition. Then the probability area of the detonation transition recording can be expressed on the coordinate plane $X$–$t$ by a ratio of the product $\Delta X \times \Delta T$ to the frame area $\delta X \times \delta T$.

The values of $\Delta X$ and $\Delta T$ follow from equation (9.18), and $\delta X$ and $\delta T$ are the frame size and time exposure, that are necessary for sharp imaging of real picture structure. Hence the probability W of the informative frame capture is as follows,

$$\mathrm{W} = \frac{\delta X \times \delta T}{\Delta X \times \Delta T}. \tag{9.19}$$

Substituting the values $\Delta X \sim 10$ cm, $\Delta T \sim 10$ μs, $\delta X = 5$ cm, $\delta T = 0.5$ μs results in probability about 2.5%, i.e., the sought recording succeeds in 2–3 cases from 100 attempts. The registration at the performance degradation leads to problems in the picture interpretation.

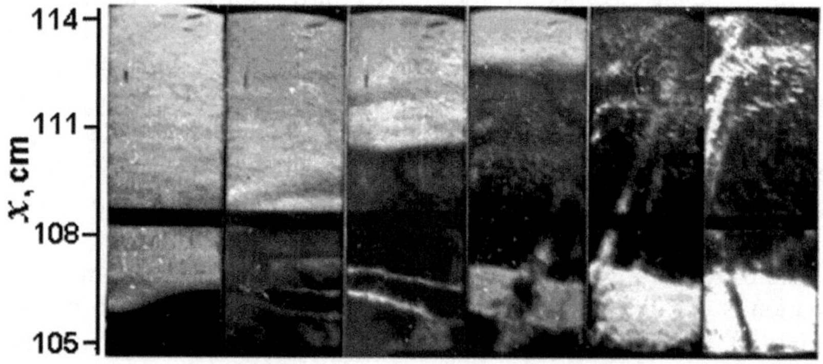


**Fig. 9.21.** Detonation transition: $t = 3.45$ ms, $\Delta t = 8$ μs ($n = 125.7 \times 10^3$ f/s), $\tau = 1$ μs

Let us consider an example. Double deceleration of the frame repetition frequency makes the observation of the detonation transition process longer, but the image resolution becomes unsuitable for the picture interpretation, as seen from the frames in Fig. 9.21. At the same time, the frame series presented in Fig. 9.21 throws light upon some other details of the phenomenon.

In particular, the frames 2–4 demonstrate interaction of the reflected shock waves moving across the pipe with the detonation wave. Also, the frames 4–6 reveal an interesting fact according to which these transverse waves do not damp during rather long time periods. All that implies that under certain conditions, for example near the detonation limits, the combustion's chemical reaction in the mixture behind the second shock wave lasts for a long time.

The reaction can maintain and even strengthen these shock waves. Then these reflected waves create the structures known as spinning detonation and pulsating detonation. Both kinds of detonation were discussed in detail by K.I. Shchelkin and J.K. Troshin [49].

Now, some valuable information on the detonation transition can be obtained from examination of the slit scans. The scanning enables us to embrace a still more prolonged part of the process and within a more extended part of the pipe. Two examples of the detonation wave origination are displayed in Fig. 9.22. Two photos have been taken at equal conditions with the only difference concerning the image visualization sensitivity.

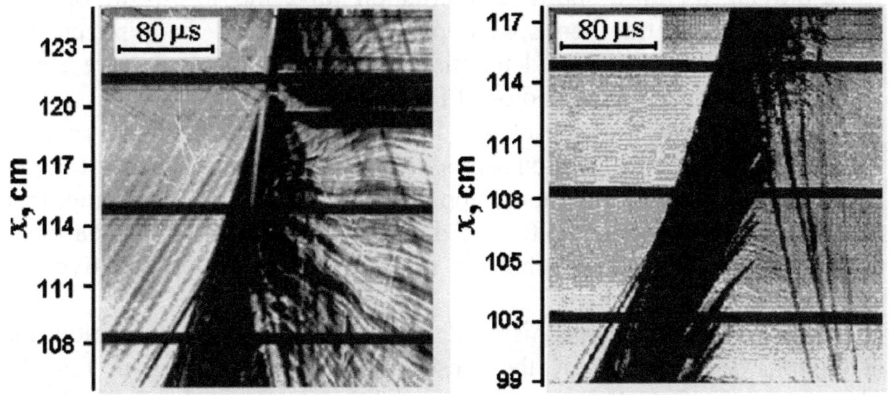


**Fig. 9.22.** Slit scans of the detonation formation process

The detonation wave is seen to arise within the second double saltus far behind the shock wave, and even behind the leading point of the combustion zone. Velocity of the detonation seems to be infinite, when it appears in the slot window field of vision. Actually, as it has been shown in Chapter 8.1, the observation means nothing else than the detonation originates on the pipe wall.

The wide horizontal dark strip on the left photo corresponds to destruction of the optical glass of the observation window. It happens at the place where the detonation wave merges with the shock wave of the second double saltus, when the pressure sharply increases in the detonation wave.

Relatively weak perturbations are visualized in the left photo owing to heightened sensitivity of the Schlieren system. However along with the improvement there is lost an opportunity to discern strong perturbations against the background of all the other perturbations. Therefore, a new reduced sensitivity has been employed to overcome the difficulty and to take the right photo in Fig. 9.22, where the weak perturbations may not be seen. So, the process development at this stage should be considered by both photos presented in the figure.

Transition to the detonation is a consequence of the persistent strengthening of the second shock wave. This strengthening happens, for the flame continues permanently emitting the compression waves which merge, catching up with each other to produce a very strong shock wave of the second double saltus. When it is just formed, the induction delay of the chemical reaction becomes so small that the combustion zone sides with the strong shock wave. The forming shock wave has the front of its shape similar to the spherical form. At one of the spots of the wave reflection from the pipe wall, the mixture temperature grows to such an extent that there arises a detonation wave reminding one of a point-wise explosion.

The detonation wave, in the shape of a hemisphere, propagates over the pipe volume in all directions. It appears on the slit scan as a detonation wave moving forward, and as a returnation wave moving in the opposite direction. The central hemisphere part moves across the pipe creating transverse oscillations. The detonation and returnation waves are clearly visible in the left and right photos of Fig. 9.22.

Comparison of the two photos illustrates the spread of the positions where the detonation transition arises. On the left photo it is about 112 cm, whereas the right photo points to approximately 117 cm. In addition, there originate more than one centers of detonation within the quasi-stationary double saltus. As the right photo in the figure shows, there are five such centers in the experiment.

The considered mechanism of the normal burning transition into detonation is confirmed by the pressure oscillograms given in Fig. 8.1. The pressure gauges have the frequency band up to the level of $10^4$ Hz. The band corresponds to traveling time of the detonation wave along the gauge surface. Therefore, the pressure pulse may be accurately measured when its duration does exceed $5 \times 10^{-5}$ s.

Certainly, the band's upper limit introduces a well-known error, when a high-frequency pulse is analyzed. However this restriction relates only to the front shape of ultra-short pressure pulses, and does not prevent measurement of the mixture state behind any shock wave in the experiments.

In conclusion, we would like to recall violent debates of the 50th years of the last century on the mechanism of the detonation transition. The respected participants, Ja.B. Zeldovich, K.I. Shchelkin and A.S. Sokolik advanced alternative opinions on the nature of the phenomenon and the principal cause of the transition. The phenomenon explanation was based on formation and development of the double saltus [50], on turbulization of the flow in front of the flame [49], and on intensification of the chemical reactions impelled by the temperature increase behind a shock wave [65].

The results considered above show that all the mentioned approaches to the problem are decisive, though each of them is appropriate to an appointed stage of the process development.

# 10
# An inverse problem of turbulence

## 10.1 Object of the inverse problem application

In previous chapters, we discussed and solved problems concerning the spatio-temporal field of turbulent flow parameter oscillations in channels for specified conditions. The velocity oscillations and their spectra displayed in Figs. 7.6 and 7.7 have revealed that the quantitative characteristic of those oscillations in time and space strongly depend on entry conditions.

This dependence enables us to make an attempt to consider the problem in the inverse formulation, with the purpose of determining parameters of the initial perturbations using the observational data on oscillation properties measured at a later time or within a neighboring spatial region of the domain. The initial perturbation properties of interest among others can be, for instance, the characteristic dimension $a$, parameter distribution over the perturbation $f(r \leq a)$, and rules of their emergence, i.e., of the sequences ***rp*** and *tp*.

The inverse problem solving procedure has proved to be rather effective for study of the optical discharge structure in the field of Besselian beams of laser radiation [94].

To prove effectiveness of the inverse problem formulation and solution, some data on physical properties of the oscillation development in a medium are necessary. Such properties are not usually measured during studies of flows in channels. But many of them are usually examined in the optical discharges produced by laser radiation beams. Therefore, we demonstrate the inverse problem method of solution by the instance of the optical discharge wherein such data are available.

As shown above, the parameter oscillation profile in a flow is formed much more rapidly than the fluid velocity is changed at moderate flows in channels. In this sense, conditions of the perturbation formation are identical in both cases to be compared. To reinforce identity of the conditions in the two cases, a peculiar laser beam is considered, which creates not one, but a series of discharges.

The wave beam optimum concentration (focusing) is known to be one of the central problems of optics, laser physics, radio physics and acoustics.



L.N. Pyatnitsky, *Turbulence Nature and the Inverse Problem*, Fluid Mechanics and Its Applications 89, 155–180.

The traditional approach to focusing is grounded on the use of devices, or any other focusing system, similar to a convex lens with spherical (or parabolic) surfaces. The device used to transform the incident radiation into a beam with a field cross-section diameter of the size

$$d \approx \frac{2.44\,\lambda}{D} f, \tag{10.1}$$

where λ the radiation wave length, $D$ means the cross-section diameter of the incident beam at the inlet of the focalizing system, is assumed to be a lens with focal length $f$.

The flat phase front of the incident wave, if any, is converted by the lens into the spherical. It becomes flat only in the focal plane, i.e., in the waist of the lens caustic. Thereat the radial distribution of the radiation intensity repeats the kind of radial profile of the incident beam. In an ideal case it is the Gaussian squared, $G^2(r)$.

The length $L_d$ of the focal area along the z axis (the longitudinal size), is one of the most important characteristics of the focused beam, which is defined by wave diffraction. The diffraction divergence angle γ and focal region length $L_d$ where the wave front is nearly flat, which is usually called Rayleigh length, are accordingly

$$\gamma \sim \frac{\lambda}{d}, \quad L_d = \frac{\pi d^2}{\lambda}. \tag{10.2}$$

To increase the radiation power concentration and augment the intensity, the waist diameter $d$ should be reduced as far as possible. However, it follows from formulas (10.2), that this reduction leads to growth of the beam divergence and the length $L_d$ shortening. Thus it is the diffraction divergence that prevents enhancement of the beam intensity and creation of thin thread-like beams at the length of $L \gg L_d$.

The diffraction divergence of the wave beam is fundamentally irremovable, nevertheless there is a way to compensate for the divergence. It follows from the diffraction divergence configuration schematically presented in Fig. 10.1. If we invert the problem formulation, and turn the wave vectors $\boldsymbol{k}(\gamma)$ in the counter direction, then the inverse wave should converge at the angle $\gamma \sim \lambda/d$ to form the waist by diameter $d$ with the wave vectors $\boldsymbol{k}_0$. An infinite set of such waists evenly distributed along the axis is expected to form a united beam at the same constant diameter.

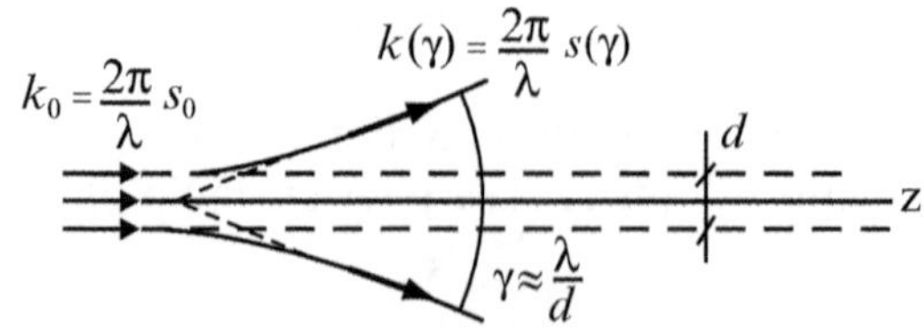


**Fig. 10.1.** Scheme of diffraction divergence

To produce such a beam, the flat front of an incident wave should be transformed into the front of conical shape WF. Some special device is necessary to realize the idea, e.g., a conical mirror or conical lens usually called an axicon [95]. An application of the axicon operation is shown in Fig. 10.2. In the figure, ***s*** is the unit vector normal to the front surface of a wave, index '0' relates to the incident beam, and index 'γ' concerns the converging wave. The subscripts of the wave vector ***k*** have similar meanings.

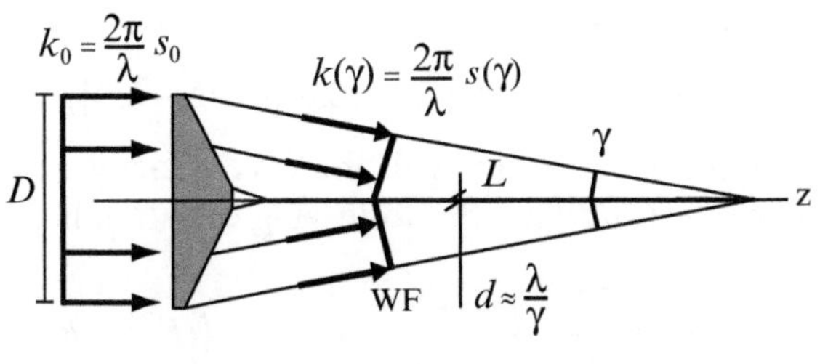


**Fig. 10.2.** Formation of extended beam of constant diameter

The length $L$ of the formed beam, unlike a Gaussian beam, depends not on the incident radiation wave length λ, but on the diameter $D$ of the incident beam and on the angle $\gamma = \alpha(N-1)$, where N is the axicon refractive index:

$$L = D/2\gamma. \tag{10.3}$$

From all that, the cylindrical symmetry of the configuration results in transformation of a Gaussian profile of the incident beam into the field's radial distribution of Besselian type $J_m(x)$ of the order $m$ in the focused beam. At the same time, the focused beam length is longer by many orders of magnitude than Rayleigh length, $L \ggg L_d$ (see review [96]).

It was expected that a threadlike plasma channel might arise in the field of such an intensive beam thanks to the axial symmetry and longitudinal uniformity of energy supply to the axicon focal volume. The first experiments were carried out with a laser radiation source. However a series of plasma channels arranged along the axis, of a non-uniform structure looking like a string of beads, appeared instead of the desired smooth plasma channel [97].

At first the "beads" had been supposed to originate because of complicated mode composition of the radiation of the laser at hand. However additional thorough examinations with the single-frequency laser [98] have yielded precisely the same outcome. Therefore, the non-uniform structure of the optical discharge in the beams with the divergence compensated has been related to some specific features of the Besselian beam itself, and to a nonlinear process of propagation of the radiation at the conical wave front [99–101].

## 10.2 Wave beam at Rayleigh divergence compensated

In the beams under consideration, similarly to Gaussian beams, a wave nonlinear propagation depends on the medium permittivity [102]

$$\varepsilon = \varepsilon_0 + +i\varepsilon'' + +\varepsilon_{\mathrm{NL}}(|E|^2),$$

where $\varepsilon''$ is the term responsible for damping, and $\varepsilon_{\mathrm{NL}}$ is a nonlinear functional responsible for wave-matter interaction and defined by the medium constitutive equations. For description of the complex amplitude

$$E(\boldsymbol{r}, t) = \mathrm{Re}\{\mathbf{e}E(r, z)\exp[-i(\omega t - kz)]\}$$

of the field strength in the beam with the divergence compensated, and at the wave number $k^2 = (\omega/c)^2\varepsilon_0$, the following equation has been used,

$$2ik\frac{\partial E}{\partial z} + \frac{1}{r}\left(r\frac{\partial E}{\partial r}\right) + \left(\frac{\omega}{c}\right)^2\left[i\varepsilon'' + \varepsilon_{\mathrm{NL}}\left(|E^2|\right)\right]E = 0. \qquad (10.4)$$

The boundary conditions defined by the volume of the radiation focused by the axicon at the aperture radius $R$ are of the form:

$$E(r, z = 0) = E_{\mathrm{in}}(r)\exp(-ikr\sin\gamma). \qquad (10.5)$$

The following requirements are accepted for the beam intensity $I_m(r)$:

$$E_{\mathrm{in}}(r > R) = 0,\ E_{\mathrm{in}}(r \le R) = I_m^{1/2}(r).$$

Linear ($\varepsilon'' = \varepsilon_{\mathrm{NL}} = 0$) solution $E^{(0)}$ of the problem (10.4, 10.5) in the z-axis vicinity ($r < z\sin\gamma$, $kr^2 < z$) at the length $\lambda/\sin^2\gamma \ll z < L$ gives the field $E^{(0)}$ distribution along the axis, which can be expressed through a Bessel function of zero index $m = 0$, $J_0(kr\sin\gamma)$, with slowly varying amplitude $E_0(z)$:

$$\left|E^{(0)}(r, z)\right| = |E_0(z)|\,J_0(kr\sin\gamma). \qquad (10.6)$$

Thorough analysis shows [100, 101], that in a nonlinear case, $\varepsilon_{NL} \neq 0$, solution of equation (10.4) at the boundary condition (10.5) leads to longitudinal modulation of the beam field $E(z)$. The modulation is of the scale $l$ which does not depend on the nonlinearity mode type:

$$L = 2\lambda/\sin^2\gamma. \qquad (10.7)$$

Photo *a* in Fig. 10.3 images a densitogram of the radial profile of the beam intensity, obtained in the experiment for the angle $\gamma = 5°$. The profile well agrees with the Bessel function of index zero. The positions $r_n$ computed for the function $J_0(r_n) = 0$ are mapped on the right side of the photo. The graph *b* represents the field longitudinal modulation, $|E|/E_0$, as a function of the ratio $z/l$ computed for four values of the parameter $E_0/E_*$.

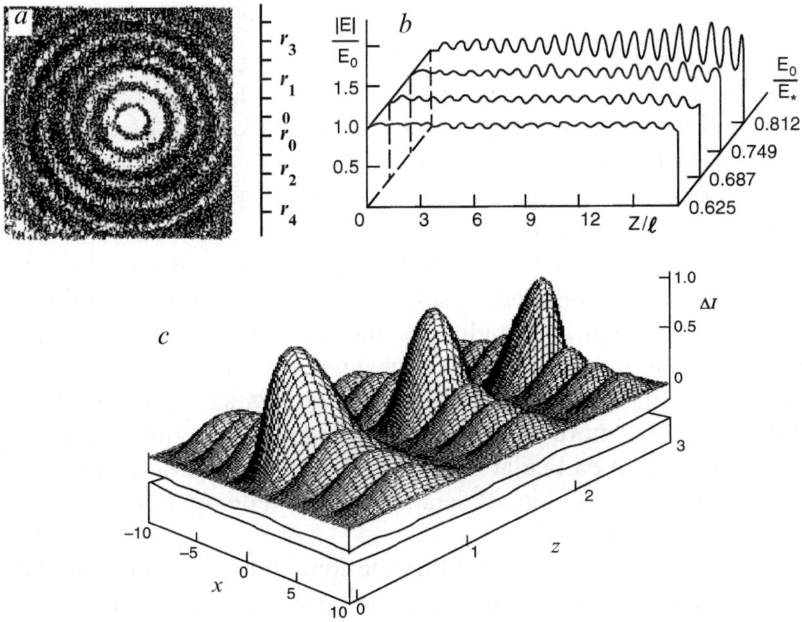


**Fig. 10.3.** Radial (*a*) and longitudinal (*b*) field distributions in the beam at $\gamma = 5°$, the field structure diagram (*c*)

The quantity $E_0$ is the linear solution of equation (10.4), and $E_*$ means the critical field value for beam self-focusing. Functions on the graph *b* are plotted as a result of the numerical solution [99] of equations (10.4), when the permittivity nonlinear part is $\varepsilon_{NL} = (n/n_{cr})\left(|E|/E_*\right)^2$. Here $n$ and $n_{cr}$ are electron density and its critical value for the frequency $\omega$.

Combination of the field radial and longitudinal distributions forms the total realistic structure of the beam intensity, of which a fragment is seen in the diagram *c* in Fig. 10.3 [94]. Herein the beam radius is presented by the dimensionless argument of a cylindrical functions, $x = kr\sin\gamma$, and the length along the z axis is expressed in scale $l$ of the longitudinal structure, $z/l$. As to intensity, it is shown as a variable part $\Delta I$ in the axial plane. The beam axis of symmetry corresponds to $x = 0$. The distance between the nearest radial neighbor maxima makes up $\delta r \approx \lambda/(2\sin\gamma)$, and the spatial period of the maximal longitudinal distribution is equal to $l = 2\lambda/(\sin\gamma)^2$.

The modulation at the period (10.7) arises, when the radiation field (10.6) is close to the critical quantity $E_*$. For example, for nonlinearity accepted in calculation of dependences shown in graph *b* in Fig. 10.3, the critical value should satisfy the following requirement:

$$0.2 \le \frac{n}{n_{cr} - n} \frac{1}{\sin^2 \gamma} \frac{|E^{(0)}|^2}{E_*^2} \le 1 . \qquad (10.8)$$

If the beam field becomes weaker, the longitudinal modulation recedes, and the modulation does not appear outside the lower limit of the requirement (10.8). Otherwise, if the field strength exceeds the upper limit, there appears an additional structure along with the basic structure (10.7), describable by the spatial period $l_1 \sim 10l$ [103].

Moreover, when the field intensity reaches the level of $5 \cdot 10^{13}$ W/cm$^2$, the phenomenon of nonlinear absorption may grow to exert influence on the channel evolution. The radiation interacts with the plasma resonance modes available in the course of channel developing. The interaction results in modulation of the plasma heating, which is followed by the appearance of one more structure of the channel, this time depending on many parameters, in particular on the medium initial pressure. This intricate process is not understood in all details. The first attempt to comprehend the process was made in [104].

It is important to emphasize that the longitudinal modulation with the spatial period (10.7) does not deface the field radial distribution which is described in a usual manner by a Bessel function. Owing to this feature, the beams with the divergence compensated have been designated as “Besselian beams”, by analogy with Gaussian beams.

This brief overview of Besselian beam properties allows us to imagine the process of structure formation in optical discharges, and to estimate the peculiarities of inverse problem applications. However at first we should consider the configurations of these structures discovered in experiments.

## 10.3 Structures of the plasma channels in lengthy wave beams

Experiments with laser radiation at the wave length of $\lambda$ = 1.06 μm focused by an axicon are described in [94, 105]. The equipment parameters were varied within wide ranges to produce Besselian beams and plasma channels of different kinds. In this way, the angle $\gamma$ was varied from 1° to 18°, that determined Besselian beam diameter in the range of $d \in [50, 3]$ μm, and the beam length $L \in [130, 1.5]$ cm. The plasma was produced in gases of ten different compositions at the pressure range 0.05–10 atm. Duration of the heating impulse lessened step by step from 50 to 0.1 ns. Accordingly the emitted energy decreased from 80 to 0.6 J.

Several methods of laser diagnostics [106] were used for thorough study of the state of the plasma channel structure. The optical perturbation distribution was visualized in the light of plasma self-radiation, as well as by Schlieren- and shadow methods. The structure state of plasma channels created by the short heating impulse at duration of 0.1 ns was investigated by means of interferometry with laser illumination at the wave length being shorter in comparison with heating radiation. Images of structural formations were recorded by means of image converters and charge-coupled devices.

Systematization of all the experimental data allowed us to discover five typical structures (with some variations) of optical discharges [94, 105, 107], which are represented in Fig. 10.4 according to chronology of their recording. On the photos, the wave front of the heating radiation propagates from left to right.

The optical discharge on the frame *a* in Fig. 10.4 is produced in air at atmospheric pressure by means of the laser impulse duration $\tau$ = 40 ns, energy $E$ = 70 J and the axicon with the angle $\gamma$ = 7.5° ($L$ = 17 cm). As seen from the photo, the discharge structure is characterized by the presence of the side branches reminiscent of petals. The angle $\beta$ of these petals concerning the symmetry axis exceeds distinctly the slope $\gamma$ of the wave vector, $\beta \gg \gamma$.

Therefore, the great disparity and origin of the petals can not be explained by the breakdown motion in direction opposite to the wave vectors of heating radiation, as is the case with the optical discharge in a focal spot of a spherical lens [108]. Note in addition, the petals take positions periodically located along the channel axis at the distance $l$ = 0.12 mm from each other, and each of them taken separately consists of small discrete sparkles of the breakdown.

The frame *b* in the figure demonstrates the discharge structure produced in argon at atmospheric pressure by a heating beam of the radiation at the same parameters as in the case of the frame *a*, i.e., $\tau$ = 40 ns, $E$ = 70 J and $\gamma$ = 7.5°. Here the large-scale structural formations of the funnel-shaped form are observed. The separate parts of them are separated by intervals of more than 1 mm.

As is known (see, e.g. [108, 109]), the threshold intensity in argon makes up $I_{th} = 1.5 \times 10^{10}$ W/cm$^2$, and the threshold in air amounts to only the value of $I_{th} = 6.5 \times 10^{10}$ W/cm$^2$, i.e., ~4.5 times more. And so the excess of the available intensity over the level of the threshold in argon is much more essential than in air, all other conditions being equal.

This seems to be the reason for the structure modification when argon as a medium-target (see frame *b* in Fig. 10.4) replaces air (frame *a* in Fig. 10.4). Nevertheless, the beads (the intensity maxima) periodically arranged along the axis are noticeable on both frames, and the distance between them, $l \approx$ 0.12 mm, corresponds to the structure period prescribed by expression (10.7).

The frame *c* has been obtained in air with a laser impulse of somewhat less duration, $\tau$ = 20 ns, with the other parameters being $E$ = 20 J, $\gamma$ = 5° ($L$ = 26 cm). Here the intensity makes up one sixth part, as compared with the case presented on frame *b*, and the specific energy is 12 times less. The periodical structure of the beads appears along the beams in this case as well, but the spacing increases up to $l \approx$ 0.28 mm. In addition, the structural formations turn into solid breakdown plasma sites, remaining in the vicinity of the symmetry axis.

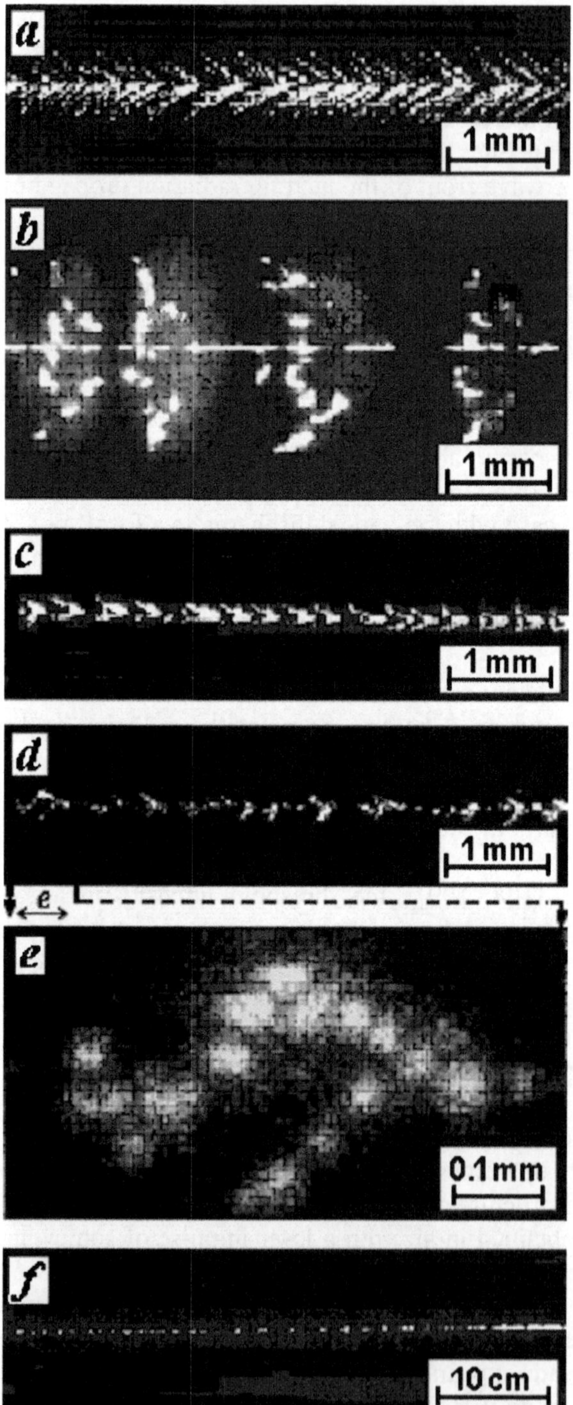


$\tau$ = 40 ns, $E$ = 70 J,
$\gamma$ = 7.5°, air 1 atm

$\tau$ = 40 ns, $E$ = 70 J,
$\gamma$ = 7.5°, argon 1 atm

$\tau$ = 20 ns, $E$ = 20 J,
$\gamma$ = 5°, air 1 atm

$\tau$ = 0.8 ns, $E$ = 17 J,
$\gamma$ = 2.5°, air 1 atm

Fragment of the frame (*d*)

$\tau$ = 0.8 ns, $E$ = 10 J,
$\gamma$ = 1°, argon 0.2 atm

**Fig. 10.4.** Typical structures of optical discharges in Besselian beams

The other photos in Fig. 10.4 were recorded by laser heating impulses of much shorter duration, $\tau = 0.8$ ns. The frame *d* is the result of an experiment in air with the set-up parameters being $E = 17$ J and $\gamma = 2.5°$ ($L = 52$ cm). As compared with frame *c*, the radiation intensity in the case of frame *d* is nearly three times as much, though the radiation specific energy amounts to only one tenth part of the previous quantity. The structure's longitudinal spacing comes to $l = 1.1$ mm. Looking at photo *d* we can conclude that the specific energy decrease deranges the shape uniformity of the discharge structural formations. However their uniform distribution remains unchangeable.

Also, each fragment of the breakdown structure consists of discrete sparkles by size of $0.02 \div 0.05$ mm, which are clustered together and form the dotted lines close to the axis. A selected fragment of frame *d* is reproduced as a magnified image in frame *e*. The photo *e* clearly shows both the discrete sparkles and how they are grouped together into lines. The slope angle $\beta$ of the lines is seen to considerably exceed the incidence angle $\gamma$ of the wave vector $\boldsymbol{k}$.

The frame *f* in Fig. 10.4 illustrates the discharge structure recorded in argon at the pressure 0.2 atm, within the beam arisen at $E = 10$ J and $\gamma = 1°$ ($\tau = 0.8$ ns). These parameters mean, that the intensity in the beam is reduced by two orders of magnitude in comparison with the previous experiment. At length $L = 130$ cm the frame *f* includes the beam part by the length of 50 cm near the axicon.

The intensity near the axicon hardly exceeds the threshold level, and amounts to about a fourth part of all the length $L$ average. The breakdowns are located along the symmetry axis in strict succession with the space period $l = 7$ mm. When moving off from the axicon, the beam intensity grows, and the separate breakdowns tend to merge, so that far off the axicon the breakdowns form the channel's continuous body.

As follows from the cited data, the longitudinal periodical structure is observed in all the typical discharges shown in Fig. 10.4. The structure spacing does not depend on gas composition, pressure, energy or on duration of the heating impulse, but only on the angle $\gamma$ of the generatrix slope of the wave front cone, shown in Fig. 10.2. Note, that other components of the structures frequently can be simultaneously seen on the frames, namely of larger and finer scales.

The structure configurations of large-scale components are evidently associated not merely with Besselian beam properties, but also with the heating impulse duration $\tau$ and with the ratio of the beam available intensity to the intensity threshold level, $I/I_{th}$, inherent to the gas used. At the same time, dependence of the fine-scale components of the structures on the experimental conditions is not so obvious. Nevertheless the fine component is present in nearly all the discharges under consideration, to a greater or lesser extent. And the discharge needs to be investigated at an early stage.

The very early stage of the optical discharge development in Besselian beams attracts special attention, for it is essentially an unexplored phenol-menon. Structures of the primary breakdowns of the discharges have been investigated first by means of recording the heating radiation scattered by the nascent plasma cells. However measurements of the cell dimensions, not to mention the parameters, have had no success in this way.

One of the reasons for failure consists in the long-lasting heating impulse. Really, even with the impulse duration of ~1 ns, the linear dimension of the breakdown body increases by ~40 μm. This augmentation may exceed the initial body size many times. Therefore, the heating impulse duration has had to be reduced ten times at least. The idea has been realized at the laboratory of Professor H. Milchberg (Maryland University, USA). The experimental facility description can be found in the references [110, 111].

Appropriate experiments [94, 105] have been carried out with the wave beam formed by the heating impulse at the parameters: $\lambda = 1.064$ μm, $\tau = 100$ ps, and $E = 0.6$ J. The energy level has been conditioned on the choice of nitrous oxide as the medium-target because of the threshold intensity being low, and of the most accessible angle $\gamma = 18°$ for concent-ration of the radiation. The beam at that angle has had diameter $2a = 2.6$ μm of the Bessel function central part, and the length $L = 1.5$ cm.

A Mach–Zehnder interferometer, illuminated by laser radiation at wave length 0.53 μm and impulse duration 70 ps, has been employed to observe the discharge development beginning from the breakdown initiation. The interferogram magnified images have been registered by means of a 512 μm matrix (320 pixels).

One point in these measurements needs to be cleared up. Besselian beam intensity attains a level of about $5 \times 10^{13}$ W/cm$^2$ in these experiments. Remember, this is the level where the absorption of the heating radiation may influence the structure of a Besselian beam. In order to avoid some unwanted distortion in interpretation of observable effects, the optical discharge has been explored in its evolution and at different pressures of nitrous oxide. Therefore, the interferograms have been registered at a pressure in the range $P \in [50, 500]$ Torr and the time interval delayed after the heating pulse maximum for within $\Delta t \in [0, 800]$ ns.

The frames *a*, *b* and *c* in Fig. 10.5 represent interferograms exposed at the instances $\Delta t = 0, 100, 250$ ns. Numbers 1 and 2 associated with the letters relate to nitrous oxide pressures $P = 200$ and $500 Topp$ (or 0.27 and 0.67 atm). The numerals on the interferogram abscissa and ordinate are the numbers of CCD matrix pixels. The space interval between neighbor pixels corresponds to 1.6 μm.

Contours of the discharge channels are clearly visible on the frame photos against the background of the interference fringes of the equal declination. The fringe shifts represent the medium local optical perturbations, which mean in fact the medium density changes.

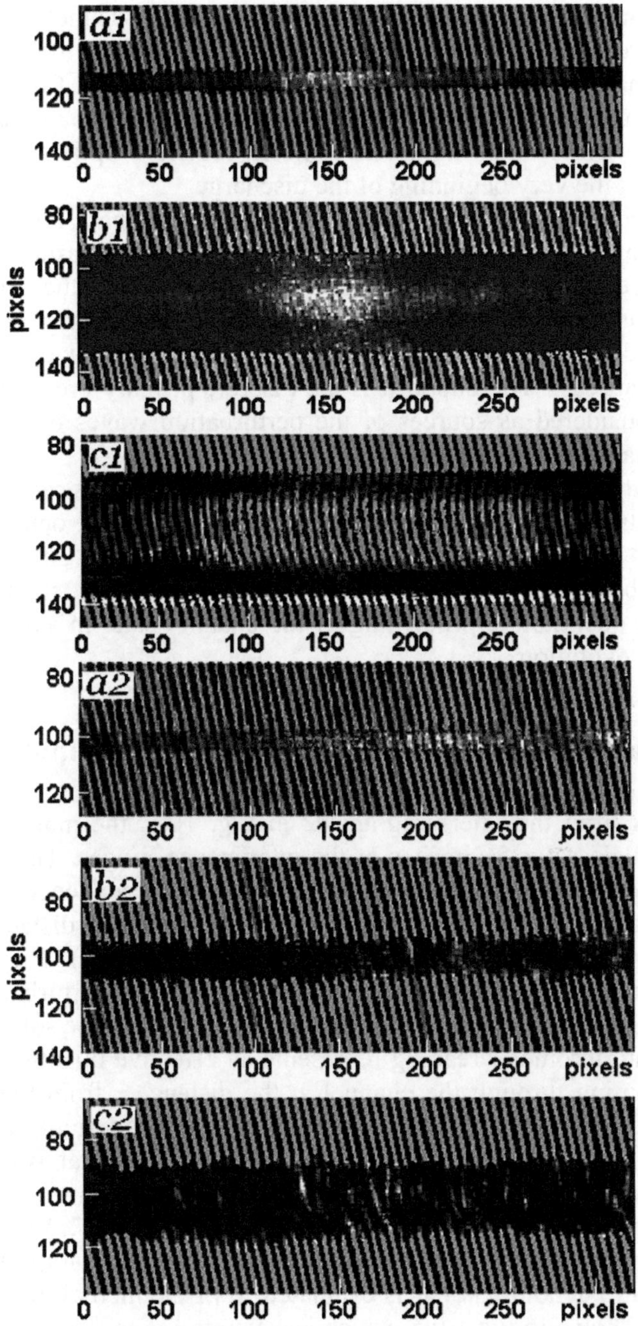


**Fig. 10.5.** Channels of optical discharges in nitrous oxide: *a1*, *b1*, *c1* ($P$ = 0.27 atm), probe impulse delay 0, 100, 250ns; *a2*, *b2*, *c2* ($P$ = 0.67 atm), probe impulse delay 0, 100, 250 ns

The interferograms $a1$, $b1$, $c1$ prove that the channel of the optical discharge remains uniform while the heating impulse is being generated, provided that the pressure is 0.27 atm. On the contrary, it follows from the interferograms numbered as $a2$, $b2$ and $c2$, that when the ambient pressure amounts to 0.67 atm, the interferometer registers optical perturbations arising from the very beginning of the discharge.

The prior problem, however, consists in a poor space resolution of the interferograms, which does not allow us to see and measure directly the primary perturbation dimensions. Notwithstanding, the fringe shift distributions within the channel images shown by the interferograms $a2$, $b2$, $c2$ testify to the discrete character of the discharge development at the first stage at least. Thereupon it is the discrete primary breakdowns which may be considered as sources of the perturbation waves observed on the channel images in Fig. 10.5.

Now, the technique developed in the previous chapters can be applied to the data extracted from these interferograms to recover the optical discharge structure at the instant of the primary breakdown. The interferogram $c2$ seems to be the most informative and therefore most suitable for the purpose of quantitative processing of the channel state in spite of the late stage of its evolution.

## 10.4 Breakdown structures in the short heating impulse

As follows from the interferogram $c2$ in Fig. 10.5, the matrix of a CCD camera contains $N = 56$ images of the interference fringes. The satisfactory resolution of the fringes enables us to describe the dependence of the fringe shift $\delta$ as a function of the coordinate $z = Nh$ ($h$ is the distance between fringes), or simply of the fringe number along $z$ axis, $\delta(N)$. The fringe shift has been mentioned to be equivalent to the gas density perturbations within the channel. Such a dependence is indeed the problem to be solved.

The interferometer reading represents an effective quantity $\delta(N)$ along the line coming through the channel at the distance $r$ from the symmetry axis, i.e., the impact parameter. So that the whole interferogram presents effective values of $\delta(N)$ as a function of $r$, and, in that way, gives an opportunity to obtain experimental dependence $\delta(N,r)$.

The graphs in Fig. 10.6 just produce the function $\delta(N,r)$ in the form of separate functions $\delta(N)$ for different values of the relative impact parameter $r/R$. The ratio denominator $R$ is evaluated from the medial diameter of the channel $c2$ shown in Fig. 10.5, which is estimated from the photo and equal to $2R = 41.0$ μm.

The top graph in Fig. 10.6 lets us see the fringe shift behavior at the distance $r/R = 5$, far off the channel axis, and in fact demonstrates the interferogram quality, or measurement accuracy at the observation conditions.

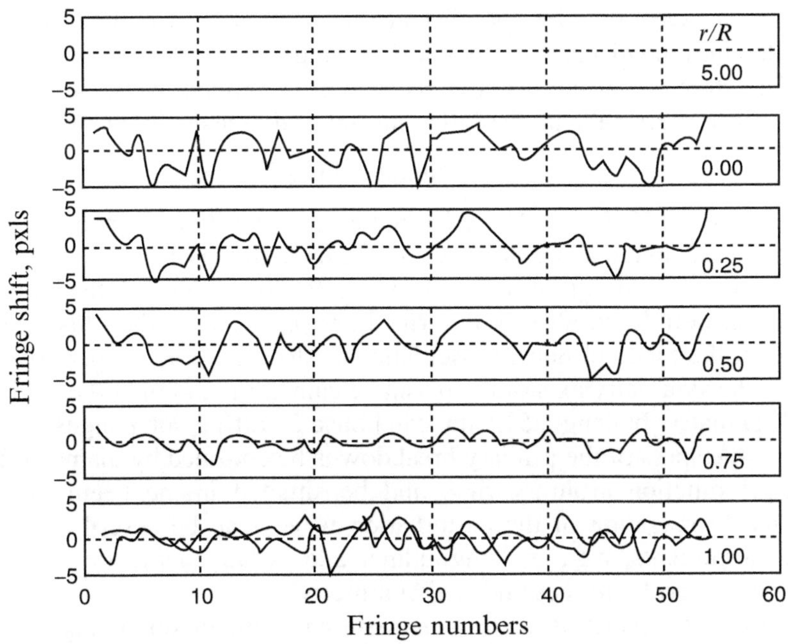


**Fig. 10.6.** Distribution of optical density $\delta(N)$ in channel *c2* (Fig. 10.5)

The next graphs describe distributions of the fringe shifts along the channel for the relative radius ratios $r/R$ = 0, 0.25, 0.50, 0.75. The graph at the bottom relates to the channel's outer surface, $r/R$ = 1, and expresses radial displacements of the surface which looks on the photo *c*2 (Fig. 10.5) to be the upper and lower boundaries.

The experimental patterns of the density perturbation distributions confirm the supposition of the discrete character of an optical discharge. The basic formation mechanism of the distributions follows from the graphs as well. Each micro-breakdown creates some perturbation in the form of a local rise in temperature and pressure. According to formula (5.2), such a perturbation propagates in space as a spherical wave packet by the thickness $2a$.

The hydrodynamic parameters vary within the packet, as the relations (5.6) predict, hence the variations are of the form $f((l_n-l))/a)$. The primary perturbations, distributed in space and in time as ***rp*** and *tp*, create in the channel a well-known to us spatio-temporal field of the density perturbations in the course of superposition of the perturbation waves.

It is natural to assume that the primary breakdowns arise in the spots whose centers are located in the beam intensity maxima. If solution (10.7) is valid and the beam structure corresponds to the image *c* shown in Fig. 10.3, then the known periodical sequence of the maxima appears at the beam axis, separated by the distance $l = 2\lambda/\sin^2$. In the case under consideration the distance is equal to $l \approx 21$ μm.

The structure in the beam cross-section is defined by Bessel function rings whose radii make up the following set of dimensionless values:

$$a = 0.063, 0.145, 0.227, \ldots, \qquad (10.9)$$

scaled by $R = 20.5$ μm. There is, however, a question, namely which of the enumerate maxima are those where the breakdowns occur.

To answer this question we can use the procedure of the turbulent oscillation definition, already described above, meaning that this time we deal with the medium density oscillations. Also, the primary perturbations arise at the symmetry axis of the plasma channel, instead of at the walls and the CCD matrix by length 512 μm can house 24 diffraction fringes.

Let the spots of the primary breakdowns be confined by diameter $2a$ of a Bessel function arbitrary ring, and be situated in the beam intensity maxima. Propagation of the perturbation waves can be described by two spatial arguments, e.g., by longitudinal and radial coordinates, $z$ and $r$, owing to the problem's cylindrical symmetry.

To get the results of the measurement and computation being formally mutually congruent, the coordinate $z$ and the fringe shift $\delta$ are convenient to express in the scale $h$. Also, the perturbation structure should be computed within the same beam length that is observed in the experiment, i.e., in the range $\{z_1, z_2\} = \{0, 55\}$. There exists one more detail to be taken into account in the problem.

When a diagnostic light beam, theoretically a ray, passes through the cylindrical channel, the interference fringe shift corresponds to the wave phase incursion along the chord at impact parameter $r_0$. We define the chord position by the coordinates $r_0 = R/2$ and $z_0 = (z_2 - z_1)/2$, for certainty. Then its length will be equal to $2r_0\mathrm{tg}\varphi$, where $\varphi$ is the azimuth angle, and position $y$ of some point along a half of the chord changes its coordinate from 0 to $r_0\mathrm{tg}\varphi$. The perturbation superposition implies adding of the perturbation waves that come to the chord length $\{z_1, z_2\}$ from all the spots $zp$ where the waves arise.

Without any limitative assumptions in the description of the perturbation wave totality, we enter the following parameters: a possible displacement $Q$ of the sequence $zp$ as a whole, $Q < 1$; an interval $q$ of casual shift of the sequence $zp$ points from the uniform distribution, $q < 0.5$; duration $t_0$ of that part of the heating impulse $\tau$ on which the breakdowns may occur; intervals of a random deviation $tp$ of the breakdown instants at the points $zp$, $tp \leq t_0$; a quantity $k$ of the primary breakdowns dispersed along the test length of the channel $\{z_1, z_2\}$ taking into account the edge effects; a function $f_{r<a}(r)$ of the density distribution within the primary disturbance; a quantity $m$ of the chord segments to sum up for getting the total effect.

Quantities $tp$, $t_0$ and $\tau$ are convenient to express in the scale of the channel development period from the maximum of the heating impulse till the moment of the structure $c2$ (Fig. 10.5) observation. This time period is known to last 250 ns. The heating impulse length amounts to $\tau = 0.4$ in this scale. Now we will write out the initial parameters of the chosen variant of the problem completely:

$$\left.\begin{array}{l} a=\{0.063,\ 0.145,\ 0.227\};\ \tau=0.4;\ t_0=0.4; \\ \{z_1, z_2\}=\{-1,\ 56\};\ \{y_1, y_2\}=\{0, r_0\mathrm{tg}(\pi/3)\}; \\ k=25;\ f(\ell)=1;\ \ m=11;\ \ Q=1;\ \ q=0. \end{array}\right\} \tag{10.10}$$

The spatial distribution of the interference fringe shifts which are equivalent to the medium density perturbations, and the spectrum appropriate to the distribution are computed by means of a program similar to the Program 6.4. Results of the computation for the perturbations with the initial sizes $a$ = 0.063, 0.145, 0.227 (the central part, the first and second rings of Bessel function) are represented in Fig. 10.7 along with the proper data derived from the experiment (frame $c2$ in Fig. 10.5).

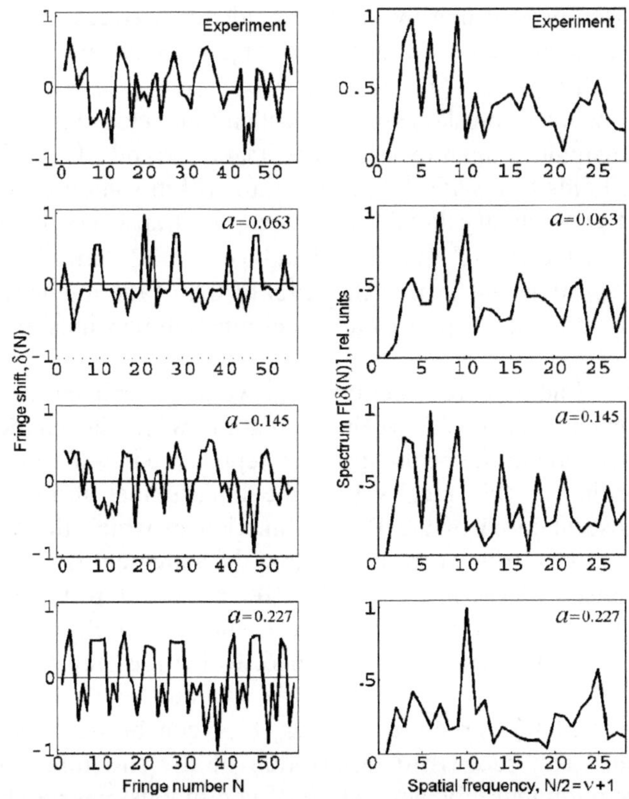


**Fig. 10.7.** Interference shifts and spatial spectra along the line $r_0/R$ = 0.5 related to experimental data and computation results for parameter $a$ = {0.063, 0.145, 0.227}

The fringe shift distributions for the indicated different conditions are assembled in the left column in Fig. 10.7. The spectra corresponding to the fringe shift distributions are placed in the column on the right. The top row represents the experimental findings.

The traits of the dependences describing the interference fringe shifts and the spectra demonstrably differ in many details. The meticulous comparison of the theoretical curves and experimental data shows that the most strict conformity with experiment demonstrates the graphs for $a$ = 0.145. When $a$ = 0.063, contour scales of the shift and the spectrum seem to be too small in comparison with data of the measurements. On the contrary, at $a$ = 0.227 the scales are too great. Therefore, further selection of the parameters (10.10) to give a more precise definition to the primary perturbations has been fulfilled with the breakdown at the dimension $a$ = 0.145.

The trial-and-error method has been used with the following parameters being varied: number of perturbations $k$, their distribution $zp$ along the $z$ axis, displacement $Q$ of all the sequence $zp$, casual interval $q$, time duration $t_0$, and sort of the function $f_{r<a}(r)$.

It has been proved that the best coincidence between the experimental and computed data occurs when the design parameters are of the following values: $k = 25$, $t_0 = 0.2$, $Q = 0.17$, $q = 0$, $f_{r<a}(r) = 1$. According to these findings, all the micro-breakdowns arise within time interval $t_0$ = 50 ps ($t_0$ = 0.2 in dimensionless quantity), and the spatial period of the disturbances, $l$ = 21 μm, holds true with the accuracy to within some percents, at least. The size of the primary breakdown does not transcend the limit of the second zero of the Bessel function at radius $a$ = 2.9 μm. The breakdown occurs in $t_0$ = 50 ps interval at the level of 0.8 of the intensity maximum. The intensity at the next rings of a Bessel function becomes insufficient for breakdown.

Thus, the findings are accepted, that we have realized the mechanism of nonlinear interaction of heating radiation with plasma development, which corresponds to the beam structure represented by the image $c$ in Fig. 10.3. According to this mechanism, the longitudinal structure discharge arises in those maxima of the Bessel function in which the field satisfies requirement (10.8). It is worth noting that the statement is valid for a short heating impulse of large intensity, $\tau$ = 100 ps, $E$ = 0.6 J. The impulse of longer duration imports additional effects into the discharge development after the first breakdowns have happened, which are seen from the frames in Fig. 10.4.

All kinds of the structures in Fig. 10.4 can be described using the above results. The successful description of the structures allows us to formulate requirements for generation of continuous plasma channels [94] that can be useful in some important applications. Also, the correct interpretation of the discharge configurations shown in Fig. 10.4 is important in one more aspect, for it can be considered as an additional argument, though indirect, in favor of the wave model of the turbulent oscillations.

## 10.5 Formation of complex structures of the plasma channel

During the heating impulse action which lasts $t_0 = 50$ ps, the primary breakdown edge covers a distance that by no means exceeds 1 μm. At this time, the breakdown body looks like a point explosion with the blast wave running in free space. However if the impulse is radiated for a longer time, the protracted radiation is absorbed by the nascent plasma and it takes the exertion of plasma pumping to influence the discharge development. Numerous facts are available on the optical discharge produced in the focal spot of spherical lenses (see, e.g. [108, 109]).

It has been ascertained that the breakdown threshold is proportional to the ionization potential, also the threshold level decreases at augmentation of the pressure, of the focal spot diameter, and of the laser impulse duration. The breakdown zone surface moves towards laser radiation at velocity up to about $10^7$ cm/s according to one of the propagation mechanisms, such as ionization wave, breakdown wave, or laser-supported detonation wave. The optical discharge propagates in various manners at different directions, for the lateral motion velocity is close to the speed of sound in the arising plasma, which is many times less compared with the mentioned quantities.

Similar phenomena should occur at radiation focusing by optical devices of the conical shape, for example by the axicon. Though Besselian beam structure (Fig. 10.3) and the discharge propagation in the traveling wave mode (running focus) should make some alterations in the discharge propagation process in this case. The sketch in Fig. 10.8 can help to imagine the picture of the breakdown front propagation along a Besselian beam.

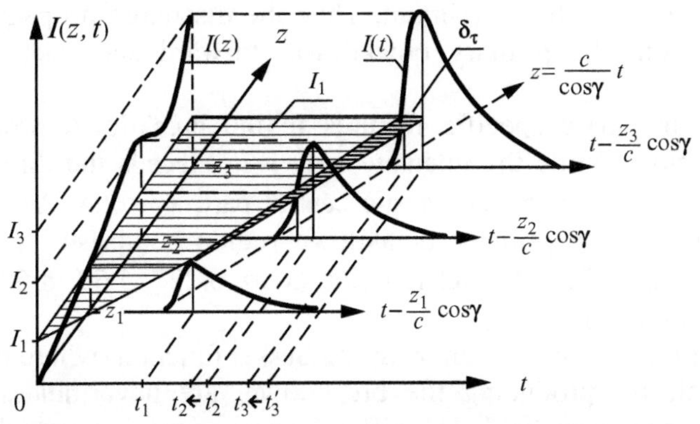


**Fig. 10.8.** A heating impulse propagation along a Besselian beam axis

Here the intensity $I(z,t)$ is presented as functions of time $t$ in the beam three sections situated at the distances $z_1$, $z_2$, $z_3$ [112, 113]. The shaded surface in the diagram designates the intensity lower level, $I_1$, whereupon nonlinear processes appear according to the relation (10.8).

While the beam axial intensity is smaller then the level of $I_1$, $I < I_1$, the intensity distribution along the z-axis monotonically increases. However, as soon as the intensity oversteps the level of $I_1$, $I > I_1$, there appears the longitudinal modulation with intensity maxima at points $z_2$ and $z_3$. The solid line $I(z)$ in the plane $z$–$I$ depicts the intensity behavior along the beam.

This delineation explains the primary breakdown enlargement with coordinate $z$ growth, and the tendency of the breakdowns to coalescence, which are observed in frame $f$ in Fig. 10.4. Behavior of the impulse intensity in time are shown by the examples at the axial points $z_1$, $z_2$ and $z_3$. Intensities attain their maxima in the instants $t_1$, $t'_2$, $t'_3$. But nonlinearity of the interaction is available only in the points $z_2$ and $z_3$, where the breakdown occurs at the instants $t_2$ and $t_3$, even before ($t'_2$, $t'_3$) the intensity maxima arrive at these points.

Thus, the primary breakdowns arise in series of the points $z_{n-1}$ and $z_n$ provided that the impulse intensity increases faster from level $I_1$ to the maximum, than it covers the distance from the point $z_{n-1}$ to the next point $z_n$, i.e., if $\Delta\tau < (z_n - z_{n-1})/v$, where $v = c/\cos\gamma$ [105]. This requirement can be written down in approximate form as

$$\Delta\tau < \frac{l\cos\gamma}{c}. \tag{10.11}$$

The breakdowns forming an optical discharge propagate in an uneven motion according to the structure consisting of the sequence of the intensity maxima of the beam. When the heating impulse lasts longer than the condition (10.11) requires, this sequence is broken. The first breakdown may arise in this case near the distance $z$, where the intensity longitudinal distribution reaches the maximum. Then the discharge propagates in both directions from the primary breakdown, forward and backward to the axicon.

There are some specific features in the breakdown propagation in radial direction. While the intensity only just exceeds the threshold level, the breakdown occurs in the caustic central part, $x_0 = 2.4$. As the impulse intensity grows, the primary breakdown region comprises new and new successive rings of the Bessel function ($x_i$ = 0.227, 5.52, 8.65, …), more distant from the axis.

With all that, some distant rings of Bessel function, where the intensity is too small for producing the breakdown, are nevertheless under the influence of the beam field. The influence may result in some abatement of the ionization potential, and in some decrease of the lower limit in the requirement (10.8). Accordingly, the heating impulse elongation leads to conditions under which the breakdown occurs at a lower level of intensity, and the nascent plasma is liable to more effective influence of the beam field. If that is the case, the breakdown should propagate in a radial direction in an uneven manner, as well as in longitudinal direction.

Let us analyze the structures produced by the impulses of nanosecond range, shown in Fig. 10.4, in reverse order, beginning from the short impulse. For this purpose it is necessary to compare the parameters characterizing the conditions at which these structures arise. To execute the idea, the data concerning the structures should be reduced to some standard which can be based, for instance, on the data taken from the experiment shown by frame *f* in Fig. 10.4.

The spots of the primary breakdowns are well perceived in the experiment with argon 0.2 atm and the impulse duration $\tau = 0.8$ ns. The spot locations are proved to be in close agreement with the axial structure of Besselian beam maxima. Also, the structure is reproduced with rather good optical resolution that has been possible thanks to the picture registration in the light of scattered radiation, and due to the long distance between the separate breakdowns, $l = 7$ mm, resulting from the very small angle $\gamma = 1°$. And though the heating impulse is almost ten times longer than the impulses related to the frames in Fig. 10.5, the remoteness of the Bessel function's first ring from the axis increases up to $\delta a = 30$ μm, i.e., approximately by the same ten times. So the observation conditions are similar in both cases.

Remember, the intensity at the initial (left) part of the beam shown in the frame *f* is of the breakdown threshold level, and its actual value corresponds to the impulse energy $E = 2.5$ J. This level can be employed as the reference value for consideration of the breakdown structures given in Fig. 10.4, destined for comparison. It is necessary to take into account in comparison, that the optical discharge threshold in argon at pressure 0.2 atm is ~ five times higher than at atmospheric pressure, and it is close to a threshold in air at 1 atm [109].

So then, the energy threshold value in air at atmospheric pressure is equivalent to $E^* = 6.5$ J for the beam parameters $\gamma = 1°$, $\tau = 0.8$ ns. It means, that longitudinal distribution of the breakdown spots under the specified conditions agrees with the maxima structure corresponding to the spacing $l$ from (10.8). As to the radial direction, the breakdown zone is limited by the central caustic radius $r_0$ of the Bessel function. Naturally, the discharge at other parameters will differ a little from that specified.

The estimation of conditions for formation of the more complex structures presented in Fig. 10.4 needs additional information concerning the ratios of the beam power density and specific energy of radiation to the proper threshold values, $w$ and $\varepsilon$. These values can be represented in the form of the proportions [94]:

$$w \sim E\gamma^3/\tau \text{ and } \varepsilon \sim E\gamma^3. \tag{10.12}$$

Let the quantities in the proportions (10.12) for the part of the discharge given by frame *f* in Fig. 10.4 be equal to unity, namely $w_f = \varepsilon_f = 1$. Then the quantities of $w_i$, $\varepsilon_i$ and the radius $r_i$ of the breakdown structure for other discharges ($i = a, b, c, d, e, f$) in the figure will be as follows:

$$\left.\begin{array}{lll} w_a = 91.2, & \varepsilon_a = 4.6\times10^4, & r_a = 0.31mm; \\ w_b = 1250, & \varepsilon_b = 6.3\times10^4, & r_b = 1.1mm; \\ w_c = 15.4, & \varepsilon_c = 386, & r_c = 0.08mm; \\ w_d = 40.6, & \varepsilon_d = 40.5, & r_d = 0.1mm; \\ w_e = w_d, & \varepsilon_e = \varepsilon_d, & r_e = r_d; \\ w_f = 1, & \varepsilon_f = 1, & r_f \approx 0.03mm. \end{array}\right\} \tag{10.13}$$

When the quantities of $w_i$ are known from the experiment, the Bessel function allows us to determine the arguments $x_i = kr_i\sin\gamma_i$ and ring numbers $n_i$ at which the radiation relative power attains the level of the breakdown. On the other hand, both the argument $x_i$ and ring number $n_i$ can be calculated by using the radius values measured from the experiments and the Bessel function. Two new designations are introduced for experimental data for the avoidance of muddle: $z_i$ instead of $x_i$, and $m_i$ instead of $n_i$. Collating the data defined by these two procedures we obtain the two lists of parameters:

$$\left.\begin{array}{ll} x_a = 59.0, & n_a = 19; \\ x_b = 803, & n_b = 251; \\ x_c = 10.0, & n_c = 3; \\ x_d = 25.7, & n_d = 8. \end{array}\right\} \left\{\begin{array}{ll} z_a = 238, & m_a = 74; \\ z_b = 850, & m_b = 266; \\ z_c = 51.7, & m_c = 16; \\ z_d = 25.8, & m_d = 8. \end{array}\right\} \tag{10.14}$$

To proceed with the inverse problem solution, the suitable values from the list (10.14) can be represented in the form of the ratios $n_i/m_i$:

$$\frac{n_a}{m_a} = \frac{19}{74}; \quad \frac{n_b}{m_b} = \frac{251}{266}; \quad \frac{n_c}{m_c} = \frac{3}{16}; \quad \frac{n_d}{m_d} = \frac{8}{8}; \tag{10.15}$$

Comparison of the data related to frames *f* and *d* in Fig. 10.4 shows that the power density in the latter case is almost 41 times as much at the same impulse duration. It is evident, that the structure *d* getting more complex is caused by the power density increase. Many micro-breakdown centers of the structure are merged into dissimilar groups, instead of the separate spots as is the case in the previous structure.

At the same time, the breakdown group centers in case of frame *d* are located along the axis at regular intervals, though the interval length amounts to $l = 1.1$ mm this time. This experimental length exactly conforms with the theoretical value at angle $\gamma = 2.5°$ available in the experiment. Also, it is important to note that both methods of estimation give the same value of the Bessel ring numbers, namely $n_d = m_d = 8$.

Meanwhile it is seen from frame *e*, a small fragment of frame *d*, that the structural group of the breakdowns resembles by sight the structure inherent to the field image plotted on diagram *c* in Fig. 10.3. Hence breakdowns of the discharge localize most likely at the points of intersection of the beam's longitudinal structure with those rings of Besselian beam maxima in which the intensity reaches the field threshold value.

The longitudinal structure *c* (Fig. 10.4) is characterized by the breakdown spots situated at the interval length $l = 0.28$ mm. This interval agrees precisely with the angle $\gamma$ rated as $\gamma = 5°$. With all this going on, the ring number of Bessel function, computed by the two above-mentioned techniques, yields quite different outcomes, $n_c = 3$ and $m_c = 16$.

The reason for such a discrepancy may, obviously, be related to the impulse duration which comes to $\tau = 20$ ns in this case, and more than 20 times exceeds the previous trial. Also, while the beam relative specific power is rather small, $w_c = 15.4$, the impulse energy $\varepsilon_c = 384$ is almost ten times more than the energy $\varepsilon_d = 40.6$. The primary breakdowns first arise evidently within the region of the three rings of the Bessel function, $n_c = 3$, at such a level of the beam power. However the wave train of heating radiation lasts $\tau = 20$ ns, i.e., the wave train has length about 6 m, and maintains the breakdown in further development during that time.

We have mentioned above, the breakdown spot produced by a spherical lens moves in the direction opposite to the wave vectors of the focused radiation at a velocity about $v \sim 10^7$ cm/s. As to other directions, the plasma thermal expansion defines the spot propagation velocity which is only $u \sim 10^6 \div 4 \times 10^6$ cm/s. The wave of ionization is proved to precede plasma development in the breakdown spot.

Generally, a front of primary breakdown produced by an axicon runs in opposite direction to the incident wave vector in the same way as in case of spherical focusing. However, unlike the spherical focusing, there exist in a Besselian beam other maxima of intensity, ready for breakdown. If such a maximum falls within the region of the ionization wave which reduces the medium dielectric strength, then a breakdown arises at the maximum as well. This process looks like a some combination of the radiation-induced (of ionization-type) and breakdown wave mechanisms, which have been observed (see, e.g. [108]) in experiments with spherical lenses.

The effect of heating impulse elongation on the breakdown propagation in a Bessel beam is illustrated by the diagram in Fig. 10.9. On the diagram are shown a side view of a piece of the beam's radial structure in the form of two rings with radii, $r_{n-1}$ and $r_n$, the vectors **u** and **v** of the discharge propagation velocities in respect to the gas in accord with two mentioned ways, and a fragment of the conical wave front WF with the wave vector $\boldsymbol{k}_0$.

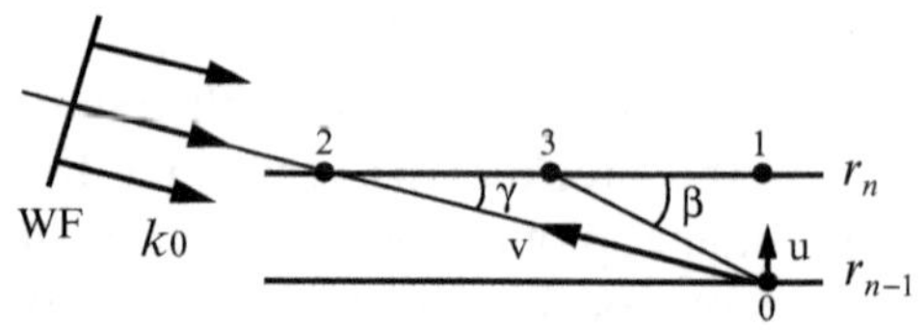


**Fig. 10.9.** Outline of breakdown propagation in a Besselian beam

Now let $s_1$ be the distance between the neighbor rings of the beam radial structure between points 0 and 1 of the hollow cylinders, and $s_2$ be the distance between points 0 and 2 of the same cylinders, but in the direction at the angle $\gamma$, $s_2 = s_1/\sin\gamma$. Velocities in these two directions amount to $V_1 = u + v\sin\gamma$ and $V_2 = v$, accordingly. Then the real direction $\beta$ of the discharge propagation can be estimated from the expression

$$\mathrm{tg}\beta = \frac{u + v\sin\gamma}{v\cos\gamma} = \mathrm{tg}\gamma + \frac{u}{v\cos\gamma}. \tag{10.16}$$

This expression reveals an important finding. The plasma within a breakdown spot being pumped by heating impulse propagates along the trajectory at the slope angle $\beta$, specified in Fig. 10.9 as line 0–3. The angle $\beta$ is seen to exceed always the angle $\gamma$.

Certainly, there have to exist a proper quantity of longitudinal maxima of the beam intensity structure (see Fig. 10.3) between the points 1 and 3 at a given radial spacing $\Delta r$ of Bessel rings. Distances between the maxima in radial, $\Delta r$, and longitudinal, $l$, directions are known from part 10.1 to depend on the heating radiation wavelength $\lambda$, argument of Bessel function $x = kr\sin\gamma$, and angle $\gamma$,

$$\Delta r = \lambda\Delta x/(2\pi\sin\gamma),\ l = 2\lambda/\sin^2\gamma.$$

Whence, the angle $\beta_m$ of the line directed from the maximum at the point 0 in Fig. 10.9 to the maximum of the next longitudinal maximum belonging to the upper radial level of the beam structure is defined by the relation

$$\mathrm{tg}\beta_m = \Delta r/l = 3.8\lambda\sin\gamma/4\pi \approx 0.3\sin\gamma. \tag{10.17}$$

Thus the requirement $\mathrm{tg}\beta_m \ll \mathrm{tg}\beta$ is satisfied in any case.

Let us consider frames *d* and *e* in Fig. 10.4 as an example to illustrate relations (10.16) and (10.17). There the angle $\gamma$ amounts to $\gamma = 2.5°$, and measurement gives the value $\beta = 13 \div 16°$. If $v = 10^7$ cm/s and $u = 2 \times 10^6$ cm/s, we have from the relation (10.16) the angle $\beta = 13.7°$. This quantity coincides with the angle $\beta = 13 \div 16°$ measured. At the same time,

$tg\beta_m \approx 0.012$ according relation (10.17), and the requirement $tg\beta_m \ll tg\beta$ is satisfied as well. This means that a breakdown in Besselian beam propagates through the intensity maximum 3 (in Fig. 10.9) of the next ring, which is located at a large distance from the point 1.

Further, when the heating impulse becomes longer and the angle $\gamma$ is still small enough, the breakdown threshold in the space between the separate plasma spots lessens. This leads to merger of the neighbor breakdowns into the plasma solid cells. For example, the picture *c* in Fig. 10.4 confirms the statement. Here the size of the breakdown spot increases during the impulse lasting $0.5\tau$ (20 ns) by $\delta r \sim 0.2$ mm, whereas separation of the rings is only $\Delta r = 0.006$ mm. The requirement $\delta r > \Delta r$ is valid in the experiment, and the inequality seems to be the reason of the intense merger process.

The structure *a* in Fig. 10.4 differs from *c* in a quantitative sense, but there is no qualitative difference between them. Really, the ring number of the radial structure computed by the beam radiation intensity makes up $n_a = 19$ ($x_a = 59.0$). At the same time, the radius measured by the breakdown spot dimension proves the number equal to $m_a \approx 74$ ($z_a = 238$).

It is clear that the structure formation goes under the influence of the same mechanisms. Though the growth of the beam's specific power and change of angle $\gamma$ from 5° to 7.5° promote increase in the number of rings involved in the breakdown process. Influence of these factors results in petal augmentation and in the real slope angle $\beta$ growth up to $\beta = 36°$.

Such a large angle value prevents from a separate breakdown merging. Hence joining up of the plasma spots into large structural formations can only be seen in the vicinity of the beam axis, where the specific power is at the highest level. Expression (10.16) gives us an opportunity to estimate the velocity ratio u/v by the angle $\beta$. In particular, for $\beta = 36°$ we obtain $u/v = 0.56$.

A peculiar type of an optical discharge is observed in the experiment shown by frame *b* in Fig. 10.4. The cells of the breakdowns look like the large bright spots arranged at the outlying periphery of the beam radial structure. Great dimensions of the spots and high electron concentration in the plasma therein shield the inner part of the beam. Therefore, the usual structure of the breakdowns at spacing $l$ (10.7) can be seen only in a close vicinity to the axis, and even in some separate parts of it. The frame *b* displays the structure as chopped fragments in the form of the side petals.

Shielding seems to be the reason for rare disposition of the breakdown groups as well. This picture evidences that the discharge formation mechanism has changed. This change is relevant to the beam parameters. Indeed, the specific power in experiment *b* is an order of magnitude more than in experiment *a*. The reduced specific power value is equal to

$w_b$ = 1,250 at the specific energy $\varepsilon_b$ = $6.3 \times 10^4$ at the same heating impulse duration. Accordingly, the intensity many times exceeds the breakdown threshold in argon at atmospheric pressure, and the breakdown has to cover the beam structure rings up to $n_b$ = 251 ($x_b$ = 803).

On the other hand, the ring number obtained by the radius $r_i$ measurement from the frame *b* is estimated as $m_b$ = 266 ($z_b$ = 851). The discrepancy of the rated and experimental data means that the intensity strongly exceeds the threshold level. To understand the fundamental peculiarity of this phenomenon, the discharge character is to be evaluated numerically. So we assess the velocity ratio u/v for the contours of structural groups.

Measuring the contour generatrix inclination results in the angle being equal to $\beta$ = 48°, i.e., the velocity ratio happens to be u/v = 1. This astonishing fact implies that the discharge propagates as a laser-supported process. Hence, the strong excess of Besselian beam intensity over the level of the breakdown threshold causes the discharge to propagate according to some new mechanism called the light detonation. Then the structure of the optical discharge within the Besselian beam degenerates into separate plasma cells united into groups of weakly predictable configuration.

The discrete character of the discharge seems to be a general property of the observable structures. While the beam intensity does not run up to ~$5 \times 10^{13}$ W/cm$^2$, the periodic longitudinal structure does not depend on the gas type or pressure. However this rule becomes inoperative as soon as the intensity exceeds the level.

It is important to emphasize that the discharge structures in Besselian beams can be discerned only until the breakdown separate spots merge. However the merging process shown in Fig. 10.4 takes place at the very beginning of the plasma channel formation. After the merging during a certain brief period, the parameters smooth in space, and the channel becomes a single whole, practically uniform.

The channel properties in the further evolution, including the diameter to which the channel eventually dilates, depend on the radiation specific energy input and its distribution along the axicon focal length. In its turn, the energy input distribution is determined by the intensity radial profile in the incident beam, and by the employed conical optical system [96].

For example, the incident beam of hyper Gaussian profile at N = 5 is focused by the axicon with the rectilinear generatrix at the angle $\gamma$ = 1°, in which the maximum energy input falls to the last quarter of the focal length. The channel photo [103] registered in the light of the plasma self-radiation is reproduced in Fig. 10.10. There have been used the laser impulse at the wavelength $\lambda$ = 1.06 μm, duration $\tau$ = 50 ns, energy $E$ = 200 J, the incident beam diameter $D$ = 4.5 cm.

**Fig. 10.10.** Plasma channel in the Besselian beam field with parameters $\lambda = 1.06$ μm, $\tau = 50$ ns, $E = 200$ J, $\gamma = 1°$, $D = 4.5$ cm, $L = 130$ cm, $d = 50$ μm

The parameters listed determine the diameter $d = 50$ μm and the length $L = 130$ cm of the Besselian beam, as well as of the initial plasma channel. The photo has been obtained by a camera at open shutter, and represents the plasma channel image integrated in time. Some increase of the channel diameter is observed at a distance of about $(3/4)L$ from the axicon.

This increase is not bound by the Besselian beam diameter, but by the longitudinal intensity distribution of the heating radiation (energy input), which depends on the initial beam intensity profile and the axicon profile. There are a number of ways to control the intensity longitudinal distribution [95, 114], which, in particular, allow us to make the intensity remain constant along the beam axis. These ways are considered in some of the author's preprints, and partly in the review [96].

It is of interest to compare the properties of other beams with those of a Gaussian beam. The Besselian beam remains nearly the same diameter along all the length $L$, whereas the Gaussian beam cross-section area would increase by ten times every 2 mm at the chosen external parameters and the same diameter. Thus the length of the plasma channel at diffraction diameter, produced by a Besselian beam, depends merely on the energy of the heating impulse.

At that, understanding of the structurization process gives an opportunity to control the discharge development and the plasma channel properties by eliminating or strengthening Besselian beam structure in response to requirements of an application. The opportunity to control the beam parameters promotes wide application of the technique [112].

It is worth paying attention to the fact that this essential peculiarity of the Besselian beam does not connect with the nature of the radiation, which may be of electromagnetic kind at any wavelength, or even of acoustic nature. In particular, when we deal with the acoustic impulse, a Besselian beam does form a waveguide similar to a tube without walls [41]. Continuous acoustic radiation may be used as a source instead of the pulse radiation. A diagram for the acoustic Besselian beam formation is plotted in Fig. 10.11.

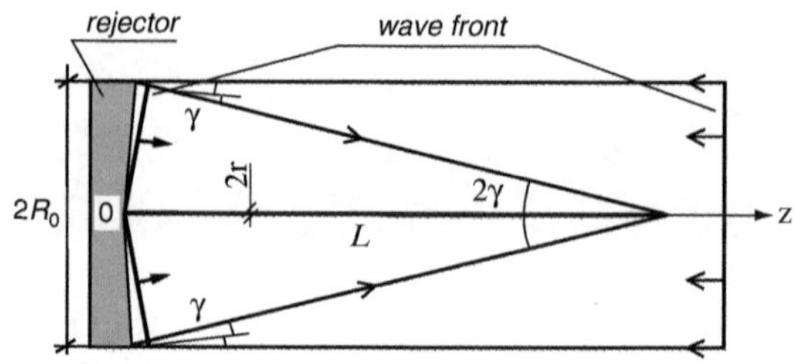


**Fig. 10.11.** Acoustic Besselian beam formation

The role of the walls in this case is played by the acoustic energy side supply in a configuration similar to the Mach reflection type. Everything written in part 10.1 relates to the field of acoustic nature. In particular, according to Fig. 10.11, the wave front propagates along the $z$ axis in the running focus regime at the velocity $c_z = c/\cos\gamma$.

It should be emphasized that the above consideration is not connected with any assumptions concerning the initial pressure or type of the medium. Therefore, the proposed procedure of focusing is not only applicable in gases, but can be employed in liquids and solids as well.

In conclusion of this section, we remind the reader that the turbulence inverse problem has been posed and successfully solved here for the first time. The wave approach to oscillations of hydrodynamic parameters employed in the solution has allowed us to interpret the intricate complex of phenomena concealed within the interaction of a laser Besselian beam with the medium in which the extended plasma channel is formed.

# Conclusion

This brief summary brings our book to a conclusion by reminding the reader of the basic principles involved in understanding the phenomenon that is called turbulence.

Turbulent flows are characterized by chaotic oscillations (fluctuations) of their hydrodynamic parameters around time-averaged values of the parameters. Widely adopted techniques for the calculation of the field of the averaged parameters are based on solving the Navier–Stokes equation. The calculated distributions of the flow parameters agree with the experimental data, at least in the cases of simple flows.

As for the oscillation field, though being of small amplitude, no unified approach to its analytical description is available, as recognized in the literature on the subject (see, e.g., §§33, 35 in [3]). The reason why the problem remains unsolved is that a definite mechanism of the fluctuations has not been identified completely since the epoch of Rayleigh, Reynolds, Lorentz, and Prandtl.

Moreover, even a physical interpretation of Reynolds' criterion lacked in clarity though numerous attempts to explain it have been made, as references [115–118] evidence it. Reynolds' hypothesis was that the energy of turbulent oscillations got converted into thermal energy in accord with an unknown law [115]. Prandtl preferred a formal theoretical manner in treating upon the problem instead of physical analysis [116]. Von Kármán resorted to the kinetic theory [117, 118]. He represented the criterion as a product of two factors – the ratios of flow velocity to sound velocity and of channel diameter to the mean free path of molecules: $\mathrm{Re} = (U/c)(d/\lambda)$.

The recent tendency is to explain the emergence of fluctuations in the main flow, which is regarded as potential, by the proliferation into it of eddies from the boundary layer where the vortex motion originates. Generally, such a proliferation contradicts Thomson's theorem, therefore they used to assume that it occurs during the boundary layer separation.

However, the separation of the boundary layer implies the emergence of a pressure gradient in the proximity of the wall. The hypothesis collides necessarily with the initial assumption, according to which the pressure is constant in every cross-section of a flow including the boundary layer, and its transverse gradient is zero. Due to this, the separation of the boundary layer is usually illustrated with the flows in which a fluid moves along an angle-shaped or curved surface. Meanwhile turbulence also emerges in the case of a flow along a plane where the above conditions are not available.

An attempt is made in the present work to consider the turbulent fluctuations to be a result of the perturbation of the flow parameters by acoustic waves. The transverse gradient of pressure in the flow is not a priori assumed to equal zero in the framework of this model. The stated formation of turbulent oscillations is examined in detail in the case of the flow in front of a piston moving in a channel.

A field of the flow velocity gradient is generated in the boundary layer due to the viscous friction near the wall. There appear in this field a vortex motion and perturbations of the hydrodynamic parameters. The perturbations are local centers of higher pressure in the zones of maximal dissipation of the energy of compression waves compensating the friction losses of the flow, when the waves are reflected from the wall. Every of the perturbations propagates in the flow as an acoustic wave packet having roughly the same size as the boundary layer thickness.

A central sector of the perturbation wave, that constitutes a main part of the wave surface, passes through the boundary layer without considerable distortion, for the path length is short. Then this part of the wave spreads to all the flow and propagates in it as an acoustic wave packet with a nearly spherical wave front. The wave packet propagating in the main, potential flow causes the parameter oscillations of common knowledge.

At the same time, a ring-shaped part of the wave adjacent to the wall does not leave the boundary layer and moves within the field of a strong gradient of the flow velocity. The wave front of the ring part of the wave packet is destroyed and the wave falls to pieces owing to refraction and diffraction in the gradient field of the boundary layer. The pieces generate therein stochastic oscillations of the flow parameters.

However the velocity drops to zero on the wall surface, so that energy of the velocity oscillations is converted to the pressure oscillations in the immediate vicinity of the wall. This wall vicinity is typically referred to as the laminar sublayer. Actually, it is not a laminar flow, for there exist the pressure oscillations.

The wave representation of the turbulence disturbances implies two mechanisms responsible for turbulent oscillations. The first one follows directly from the fact that hydrodynamic parameters within every wave packet oscillate. Superposition of the propagating wave packets, as well as of those reflected by the walls, forms the spatio-temporal oscillations field.

The second mechanism is similar to the generally recognized one, though somewhat corrected. The wave packets entrain some portions of the fluid from the boundary layer into the main flow. The vorticity frozen in the fluid is carried along with the fluid portions into the main flow.

Diffusion and fading of eddies within the flow core create a disordered field of perturbations similar to turbulent oscillations.

The wave model of turbulence provides a simple physical explanation of such phenomena as the synchronism between the velocity and pressure oscillations, the kind of distribution of the oscillations in the channel cross-section, the dependence of the transition from laminar to turbulent flow state on an incoming flow, the intermittency of turbulence, the influence of the wall roughness, the non-dissipative transfer of energy from large-scale to small-scale oscillations, and the dissipation of energy in them. The chaotization of oscillations is interpreted naturally in the framework of the proposed approach as well.

The principles on which the wave representation of turbulence is based are confirmed experimentally. The clearly formulated model of the phenomenon makes it possible to express analytically the spatio-temporal field of turbulent oscillations. As an example of the model application, the transition from normal combustion to detonation in pipes is considered. In particular, it is shown that the process involves the formation of a turbulent flow in front of the flame, which specifies the process further development.

Simulations show that the spatio-temporal field of oscillations and especially their spectra depend strongly on the problem's initial conditions. This circumstance makes it possible to set and solve the corresponding inverse problem.

This opportunity has been used and applied to study the optical discharge produced by a Besselian beam of laser radiation. In such studies, difficulties arise since the subnanosecond durations and the small size of the initial breakdown zones make it impossible to investigate experimentally the most important initial stage of the discharge. Solving the inverse problem for the perturbations structure recorded during the subsequent stages has allowed to identify the time instants and locations of the initial breakdowns, and their sizes as well.

# References


1. Reynolds O. On the experimental investigation of the circumstances which determine whether the motion of water shall be direct or sinuous, and the law of resistance in parallel channels. Phil. Trans. Roy. Soc. London. V.174, pp.935–982 (1883).
2. Schlichting H. Entstehung der Turbulenz. Handbuch der Physik, b. VIII/1. Strömungsmechanik I. Berlin-Göttingen-Heidelberg (1959).
3. Landau L., Lifshitz E. Course of theoretical physics. V.6. Fluid Mechanics. Oxford. Pergamon. Repr (1982).
4. Reynolds O. On the dynamic theory of incompressible viscous fluids and determination of the criterion. Phil. Trans. Roy. Soc. V.186A, pp.123–164 (1895).
5. Strutt J.W. (Lord Rayleigh). On the instability of certain fluid motions. Proc. Lond. Math. Soc. V.11, pp.57–71 (1880); V.19, pp.67–80 (1887).
6. Lorentz H.A. Zittingsverlag Akad. Amsterdam. V.6, p.28 (1897); Űber die Entstehung turbulenter Flűsigkeitsbewegungen und űber den Einfluβ dieser Bewegungen bei der Strőmung durch Rőhren. Abh. Theor. Physik., Teubner, Leipzig. s.43 (1907).
7. Prandtl L. Über Flüssigkeitsbewegung bei sehr kleiner Reibung. Verhandl. III. Intern. Math. Kongr. Heidelberg. pp.484–491 (1904).
8. Schlichting H. Grenzschicht-Theorie. V.G. Braun. Karlsruhe (1960).
9. Prandtl L. Bemerkungen zur Teorie der freien Turbulenz. ZAMM. V.22, ss.241–243 (1942).
10. von Kármán Th. Progress in statistical theory of turbulence. Proc. Natl. Acad. Sci. Washington. V.34, pp.530–539 (1948). (See also: Selected works of Von Kármán Theodore, volumes 1–4. 1956. [SI]: Butterworths.)
11. Richardson L.F. Weather prediction by numerical process. Cambridge University Press (1922).
12. Monin A.S., Yaglom A.N. Statistical Fluid Mechanics. Cambridge University Press. V.I, 1997; V.II (1998).
13. Taylor G.I. Statistical theory of turbulence. Proc. Roy. Soc. London A, v.151. Parts 1–4, pp.421–478 (1935), Part 5, pp.421–478 (1935); Correlation measurements in a turbulent flow through a pipe. Proc. Roy. Soc. A. V.157, pp.537–546 (1936).
14. Колмогоров А.Н. Локальная структура турбулентности в несжимаемой жидкости при очень больших числах Рейнольдса. ДАН СССР. Т.30. No.4, cc.299–103 (1941); Рассеяние энергии при локально изотропной

турбулентности в несжимаемой жидкости. Изв. АН СССР. Т.32, № 1. сс.19–21 (1941). (See reviews: Friedlander S.K., Topper L. Turbulence, classic papers on statistical turbulence. New York. Interscience (1961); Goering H. Sammelband zur statistischen Theorie der Turbulenz. Berlin. Akademie-Verlag (1958)).

15. Обухов А.М. О распределении энергии в спектре турбулентного потока. Известия. АН СССР, Геофиз. Т.5, No.4–5, сс.453–466 (1941); Oboukhov A.M. On invariant characteristics of systems of fluid mechanical type. Fluid Dynam. Trans. V.5, No.2, pp.193–199 (1971). Oboukhov A.M., Dolzhansky E.V. On simple models for simulation of nonlinear processes in convection and turbulence. Geoph. Fluid Dyn. V.6, pp.195–209 (1975).
16. Batchelor G.K. An Introduction to Fluid Dynamics. Cambridge University Press (1967).
17. Frish U. Turbulence: The Legacy of A.N. Kolmogorov. Cambridge University Press (1995).
18. Tritton D.J. Physical Fluid Dynamics, 2nd ed. Oxford. Clarendon (1998).
19. Davidson P.A. Turbulence, An Introduction for Scientists and Engineers. Oxford University Press (2004).
20. Shaughnessy E.J. Introduction to Fluid Mechanics. Oxford University Press (2004).
21. Banerjee S., Eaton J.K., Ed. Turbulence and Shear Flows Phenomena. First International Symposium. New York. Begell House (1999).
22. Ruelle D., Takens F. On the nature of turbulence. Comm. Math. Phys., v.20, pp.167–192 (1971).
23. Ruelle D. Strange attractors as a mathematical explanation of turbulence. Lecture Notes in Physics, Statistical Models and Turbulence. V.12, pp.292–316 (1975).
24. Landa P.S. Turbulence in nonclosed fluid flows as a noise-induced phase transition. Europhys. Lett. v.36, pp.401–406 (1996).
25. Landa P.S., McClintock P.V.E. Development of turbulence in subsonic submerged jets. Phys. Rep. v.397, pp.1–62 (2004).
26. Ginevsky A.S., Vlasov Ye.V., Karavosov R.K. Acoustic control of turbulence jets. Heidelberg, Germany. Springer. (2004).
27. Landa P.S., Zaikin A.A., Ginevsky A.S., Vlasov Ye.V. Turbulence and coherent structures in subsonic submerged jets. Control of the turbulence. Int. J. Bifurcat. Chaos. V.9, No.2, pp.397–414 (1999).
28. Landa P.S., Zaikin A.A., Rosenblum M.G., Kurths J. Control of noise-induced oscillations of pendulum with randomly vibrated suspension axis. Phys. Rev. E, v.56, pp.1465–1470 (1997).
29. Landau L.D. On the turbulence problem (in Russian). Soviet Physics-Doklady. V.44, No.8, cc.339–342 (1944).
30. Hopf E. Statistical hydrodynamics and functional calculus. J. Rat. Mech. Anal. V.1, No.1, pp.87–123 (1952).
31. Pyatnitskiĭ L.N. On the nature of turbulence. J. Exp Theor. Phys. V.86, No.1, pp.107–114 (1998).
32. Pyatnitsky L.N. The spatio-temporal field of turbulent pulsation. Tech. Phys. Lett. V.29, No.12, pp.1044–1047 (2003). springeronline: http://dx.doi.org/10.1134/1.1639469.

33. Pyatnitsky L.N. The structure of a turbulent boundary layer. Technical Physics Letters. V.29, No.5, pp.357–360 (2003). springeronline: http://dx.doi.org/10.1134/1.1579793.
34. Pyatnitsky L.N. Turbulent flow boundary layer. Proceedings of 5th International Workshop on Magneto-Plasma Aerodynamics in Aerospace Applications. Moscow. IVTAN. 2003. pp.335–344.
35. Pyatnitsky L.N. Wave model of hydrodynamic turbulence. Applied Physics (Moscow, in Russian). No.6, pp.5–23 (2003).
36. Pyatnitsky L.N. Mechanism of turbulent pulsation in channels. Tech. Phys. V.49, No.2, pp.194–202 (2004). springeronline: http://dx.doi.org/10.1134/1.1648955.
37. Pyatnitsky L.N. Acoustic Perturbations and Turbulent Pulsations. Tech. Phys., V.52, No.6, pp.799–802 (2007).
38. Lighthill J. Waves in fluids. Cambridge University Press (1979); Waves in Fluids and Gases. Leeds University Press (1965).
39. Landau L., Lifshitz E. Course of theoretical physics. V.2. The Classical Theory of Fields. Oxford. Pergamon. Repr (1982).
40. Nikuradse J. Gesetzmässigkeiten der turbulenten Strömung in glatten Rohren, Forschungheit No.356. Beilage zu Forschung auf dem Geleiete des Ingenieur-Wesens (1932).
41. Pyatnitskii L.N. Acoustic pulse in a wall-less waveguide. Tech. Phys. Lett. V.28, No.3, pp.246–249 (2002); springeronline: http://dx.doi.org/10.1134/1.1467290.
42. Smaryshev M.D., Dobrovolskii Yu.Yu. Hydroacoustic antennas. Leningrad (1984) (in Russian).
43. Sedov L. Similarity and dimensional methods in mechanics. London. Infosearch (1982).
44. Pyatnitsky L.N. Mechanism for hydrodynamic pulsations. Physics of Vibrations. Moscow. V.8, No.3, pp.185–207 (2000).
45. Pyatnitsky L.N. Simulation of turbulence. J. Exp Theor. Phys. V.92, No.4, pp.576–593 (2001).
46. Pyatnitsky L.N. A concept of turbulence. Proceedings of 3rd International Workshop on Magneto-Plasma Aerodynamics in Aerospace Applications. M. IVTAN. pp.407–414 (2001).
47. Pyatnitsky L.N. Origin of hydrodynamic pulsations. Proceedings of 4th International Workshop on Magneto-Plasma Aerodyn. in Aerospace Applications. Moscow. IVTAN. 2002. pp.340–348.
48. Hugoniot H. Sur la propagation du mouvement dans les corps et spécialement dans les gaz parfaits. Journ. de l'école poltecknique. No.57, p.3 (1887); No.58, pp.1–125 (1889).
49. Shchelkin K.I., Troshin Ja.K. Gas dynamics of combustion. Moscow. Academy of Sciences of USSA. 255 p. (1953); Mathematical theory of combustion and explosion. Moscow. Nauka. 270 p. (1980).
50. Zeldovich Ja.B. Theory of combustion and detonation in gases (in Russian). Moscow-Leningrad. Academy of Sciences of USSA. 125p. (1946).
51. Pyatnitskii L.N. Flame acceleration mechanism in the transition of normal combustion to detonation. Soviet Physics-Doklady. V.7, pp.495–498 (1962).

52. Pyatnitskii L.N. On structure of dual simple wave at non-stationary burning. Soviet Physics-Doklady (in Russian). V.176, pp. 631–634 (1967).
53. Pyatnitskii L.N. State of combustion gas determined by shock wave parameters (in Russian). J. Eng. Phys. Minsk. V.5, pp.21–29 (1962).
54. Markstein G. Instability phenomena in combustion waves. Fourth International Symposium on Combustion. Baltimore, MD. Williams/Wilkins (1953). pp.14–32; Markstein G., Somerse L. Cellular flames structure and vibratory flame movement in n-butan-metan mixtures. Ibid. pp.369–375.
55. Blackshear P. Driving standing waves by heat addition. Ibid. pp.381–393.
56. Courant R., Friedrichs K.O. Supersonic flow and shock waves. New York. Interscience. 425 p. (1948).
57. Liepman H.W., Roshko A., Dhawan S. On reflection of shock waves from boundary layers. NACA Rep. 1100 (1952).
58. Bogdanoff S.M., Kepler C.E. Separation of a supersonic turbulent boundary layer. JAS. V.22, pp.414–424 (1955).
59. Reichardt H. Messungen turbulenter Schwankungen. Naturwissenschaften. V.26, No.24/25, pp.404–408 (1938).
60. Klebanoff P.S. Characteristics of turbulence in boundary layer with zero pressure gradient. NACA. Rep.1247 (1955).
61. Berthelot M., Vieille P. Compt. Rend. V.93, p.18 (1881); Ann. Chim. Phys. V.28, p.289 (1883).
62. Mallard E., Le Chatelier H. Compt. Rend. V.93, p.145 (1881); Recherches expérimentales et théoriques sur la combustion des mélanges gazeux explosives. Paris (1883).
63. Mikelson V.A. On normal velocity of detonating gases inflammation (in Russian). Collected works, v.1. Moscow. Moscow University (1930).
64. Dixon H.B. Phil. Trans. Royal Soc., A, v.200, p.315 (1903).
65. Sokolik A.S. Combustion and detonation in gases. Moscow-Leningrad. USSR Academy of Sciences. 155 p. (1934); Spontaneous ignition, flame and detonation in gases. Moscow. USSR Academy of Sciences. 210 p. (1960).
66. Lewis B., von Elbe G. Combustion, flames and explosions in gases. New York Academy Press. 465 p. (1951).
67. Jost W. Explosion and combustion processes in gases. Baltimore, MD. Williams & Wilkins (1949).
68. Hitrin L.N. Physics of combustion and explosion (in Russian). Moscow University. 452 p. (1957).
69. Chapman D. Phil. Magazine. V.47, p. 90 (1899).
70. Jouguet E. Mécanique des explosives. Doin et Fils. Davis, France (1917).
71. Oppenheim A. Gasdynamic analysis of the development of gaseous detonation and its hydrodynamic analogy. Fourth International Symposium on Combustion. Baltimore, MD. Williams & Wilkins. pp.411–419 (1953).
72. Adams G., Pack D. Seventh International Symposium on Combustion. London. p.812 (1955).
73. Boa-Teh Chu. On the generation of pressure waves at plane flame front. Fourth International Symposium on Combustion. Baltimore, MD. Williams & Wilkins. pp.411–419 (1953).

74. Zeldovich Ja.B. Companeetz A.S. Detonation theory (in Russian). Moscow. Physmathlit. 275 p. (1955).
75. Kirsten H. Experimentelle Untersuchengen der Entwicklung der Geschwindigkeitsverteilung der turbulenten Rohrströmung. Leipzig (1927).
76. Glass J.J., Martin F.J. Appl. Phys. V.26, p.113 (1955).
77. Martin F., White D. Seventh Intern. Symposium on Combustion. London. p.856 (1959).
78. von Kármán Th. Some aspects of theory of turbulent motion. Proceedings of International Congress for applied mechanics. Cambridge (1934).
79. Wohl K., Shore L., von Rosenberg H., Weil C. Fourth International Symposium on Combustion. Baltimore, MD. Williams & Wilkins. pp. 425–439 (1953).
80. Karlovitz B. Flame in turbulent flow. Seventh International Symposium on Combustion. Lnd. pp.604–608 (1958).
81. Egerton A., Saunders O., Lefebvre A., Moore N. Some observations by Schlieren technique of the propagation of flames in a closed vessel. Fourth International Symposium on Combustion. Baltimore, MD. Williams & Wilkins. pp.289–294 (1953).
82. Guenoche H., Jouy M. Changes in the shape of flames propagating in tubes. Fourth International Symposium on Combustion. Baltimore, MD. Williams & Wilkins. pp.294–297 (1953).
83. Schmidt E., Steinicke H., Noubert U. Flame and schlieren photographs of combustion waves in tubes. Fourth International Symposium on Combustion. Baltimore, MD. Williams & Wilkins. pp.459–465 (1953).
84. Dryden H.L. Review of published data on the effect of roughness on transition from laminar to turbulent flow. J. Aeron. Sci. V.20, pp.477–482 (1953); Some recent contributions to the study of transition and turbulent boundary layers. Sixth International Congress for Applied Mechanics. Paris, Sept. (1946).
85. Lin C. The theory of hydrodynamic stability. Cambridge. Cambridge University Press (1955).
86. Kutateladze S.S. Parietal turbulence (in Russian). Novosibirsk. Nauka. 228 p. (1973).
87. Becker R. Z. Elektrochem. V.42, pp.457–463 (1936).
88. Dagger G., Graab D. Flames velocities of hydrocarbon-oxygen-nitrogen mixtures. Fourth International Symposium on Combustion. Baltimore, MD. Williams & Wilkins. pp 209–216 (1953); Dagger G.L. J. Amer. Chem. Soc. V.72, pp.5271–5276 (1950).
89. Novogilov B.V. Cain and thermal flames. Moscow. Nauka (1980); Mathematical theory of combustion and explosion (in Russian). Moscow. Nauka (1980).
90. Serbinov A.I. Soviet Physics-Doklady. V.129, pp.627–671 (1959).
91. Serbinov A.I., Troshin J.K., Shchelkin K.I. Soviet Physics-Doklady. V.145, pp.1314–1317 (1962).
92. Semenov N.N. Chain Reactions. Leningrad. Goshimizdat. 256 p. (1934).
93. Frank-Kameneckij D.A. Diffusion and heat transfer in chemical Kinetics (in Russian). 3d ed. Moscow. Id-intellect. 408 p. (1987).
94. Pyatnitsky L.N. Structures of the plasma channel in Besselian beam. Plasma Phys. Rep. V.27, No.9, pp.799–812 (2001).

95. Pyatnitsky L.N. et al. Device for EM Besselian beam formation in transparent medium. Patented invention (19)SU(11) 1753446 A1, Bulletin for discoveries and inventions, USSR State Committee on Discoveries and Inventions, No.29, 184 (1992), see also: Rospatent, Federal Institute of Industrial Property, www.rupat_old
96. Pyatnitsky L.N., Korobkin V.V. Wave beams with diffraction compensation and extended plasma channels. Proceedings of Prochorov's Institute of General Physics, RAS. V.57, pp.59–111 (2000).
97. Bunkin F.V., Korobkin V.V., Polonskii L.Y., Pyatnitskii L.N. Laser spark at continuous channel in air (in Russian). Quantum Electronics. V.10, No.2, pp.443–444 (1983).
98. Kamushkin A.A., Klinkov V.K., Korobkin V.V., Margolin L.Y., Polonskii L.Y., Pyatnitskii L.N. Air breakdown with single-frequency laser radiation focused by axicon. "Kratkie soobshchenia po fizike" of P.N. Lebedev Physics Institute, Russian Academy of sciences. No.11, pp.40–42 (1988).
99. Andreev N.E., Batenun V.M., Margolin L.Y., Polonskii L.Y., Pyatnitskii L.N., Aristov J.A., Zykov A.I., Terterov N.M. Self-modulation effect in raylike laser beams (in Russian). Tech. Phys. Lett. V.15, No.3, pp.83–87 (1989).
100. Andreev N.E., Aristov Y.A., Polonskii L.Y., Pyatnitskii L.N. Bessel beams of electromagnetic waves: self-action and nonlinear structures. Sov. Phys. JETP. V.73 (6), pp.969–978 (1991).
101. Andreev N.E., Margolin L.Y., Pleshanov I.V., Pyatnitskii L.N. Hollow beams of electromagnetic waves: formation and nonlinear propagation in plasma. JETP. V.78 (5), pp.663–672 (1994).
102. Lugovoi V.N., Prokhorov A.M. Theory of intense laser radiation propagation in nonlinear medium. Soviet Physics–Uspekhi. V.T.111, No.2, pp.203–247 (1973).
103 Pyatnitsky L.N. Structure of extended laser spark. Proceedings of Fourth International Workshop on Magneto-Plasma Aerodynamics in Aerospace Applications. Moscow. IVTAN. pp.118–128 (2003).
104. Cooley J., Antonsen T. (IREAP, USA); Milchberg H., Fan J., Parra E. (IPST, USA); Margolin L., Pyatnitskii L. (IHT RAS, RF). Parametric Instability in the Formation of Plasma Waveguides. Phys. Rev. E. V.62, No. 6, pp.R7603–R7606 (2000).
105. Pyatnitsky L.N. Inhomogenieties of a laser-induced spark in wave beams at the diffraction divergence compensated (in Russian). Applied Physics. No.1, pp.55–64 (2003).
106. Pyatnitsky L.N. Laser diagnostics of plasmas. Plasma Technology: Fundamentals and Applications. Ed. M. Capitelli and C. Gorse. New York. Plenum. pp.11–27 (1992); Plasma laser diagnostics (in Russian). Moscow. Atomizdat. 424 p. (1976).
107. Pyatnitsky L.N. Structures of extended laser spark. Proceedings of 3rd International Workshop on Magneto-Plasma Aerodynamics in Aero-Space Applications. Moscow. IVTAN. pp.407–414 (2001).
108. Raizer Yu.P. Breakdown and heating of gases under the influence of a laser beam. Uspekhi Fizicheskikh Nauk, V.8, pp.650–685 (1966); Optical discharges. Uspekhi Fizicheskikh Nauk, V.23, pp.789–814 (1980); Лазерная искра и распространение разрядов. М. 1974; Физика газового разряда. Наука. М. 1987. 591с.

109. Ostrovskaya G.V., Zaidel A.N. The laser spark in gases. Uspekhi Fizicheskikh Nauk. V.111, No. 4, pp.579–615 (1973).
110. Clark R., Milchberg H.M. Phys. Laser-driven implosion of a cylindrical plasma. Phys. Rev. E. V.57, p.3417 (1998).
111. Fan J., Clark T.R., Milchberg H.M. Generation of plasma wave-guide in elongated, high repetition rate gas jet. Appl. Phys. Letters. V.73, No.21, pp.3064–3067 (1998).
112. Bychkov S., Marin M., Pyatnitsky L. X-ray medium on the base of long optical discharge. Proceedings of International Colloquium on X-ray Lasers. Schliersee, Germany. IOP Publishing. pp.439–443 (1992).
113. Bychkov S.S., Marin M.Ju., Pyatnitsky L.N. Extensive laser discharge (in Russian). Proceedings of Prochorov's Institute of General Physics, RAS. V.50, pp.166–177 (1995).
114. Pyatnitsky L.N. et al. Device for laser spark production. Patented invention (19)SU(11) 1082292 A, Bulletin for discoveries and inventions, USSR State Committee on Discoveries and Inventions, No.39, 189 (1984), see also: Rospatent, Federal Institute of Industrial Property, www.rupat_old
115. Reynolds O. On the Dynamical theory of incompressible viscous fluid and determination of the Criterion. Phil. Trans.Roy. Soc (1895).
116. Prandtl L. Neuere Ergebuisse der Turbulenziorsehung. V.D.I. V.77, No. 5 (1933).
117. von Kármán Th. Gastheoretische Deutung der Reynoldsshen Kenuzahl Abhandungen aus dem Aerodynamischen Institut Aachen. No.4 (1925).
118. von Kármán Th. Uber die Stabilitat der Laminarstroming und die Theorie der Turbulenz. Abhandlungen aus dem Aerodynamishen Institut Aachen. No.5 (1925).

# Index


**A**

Acoustic wave packet 7, 13, 23, 46, 52, 73, 84, 87, 102, 107, 114, 118, 124, 129, 150, 151, 179, 182
- cylindrical 116, 118, 157, 159, 168, 176
- direct 52, 73, 119
- emission 13, 52, 113, 116, 126, 130, 139, 141, 144, 150, 151, 153, 160
- origin 4, 7, 10, 27, 51, 55, 56, 58, 61, 62, 54, 73, 74, 77, 82, 90, 93–101, 103, 107, 108, 110, 113, 118, 119, 122, 123, 134, 157, 161, 166, 168, 182
- plane 3, 48, 49, 110, 114, 139, 156
- propagation in boundary layer 70, 113, 119
- propagation in moving fluid 58
- propagation in channels 49, 58, 59, 61, 74, 76, 85, 87, 93–95, 102
- propagation in flow 14, 26, 55, 56, 59, 66, 68, 73, 87, 116, 117
- propagation in pipes 76, 81, 82, 83, 132, 134, 150, 151
- propagation inner problem 123
- propagation outer problem 123
- ray 70, 168
- reflected 73, 118
- spherical 13, 14, 26, 27, 49–53, 55, 56, 58, 60, 70, 78, 86, 103, 104, 109, 110, 113, 114, 116–118, 130, 133, 151, 154–156, 161, 167, 171, 173, 182, 183

Acoustic perturbation 7–14, 23–27, 33, 46, 51–53, 55–61, 63, 64, 66–88, 91–101, 103–105, 107, 108, 111–114, 116, 118–123, 127, 129–131, 136, 155, 160, 164, 168–170, 182, 184
- primary 2, 7, 13, 24, 69, 113, 116, 124
- primary harmonic 1, 2, 8–11

Acoustic wave superposition 4, 13, 14, 24, 47, 56, 78, 79, 85, 100–102, 104, 107, 113, 118–120, 123, 167, 168, 183

Adhesion phenomenon 2, 21, 40

Adiabatic fluid motion 18, 23, 44, 126

Approach to turbulence onset 2, 4, 5, 7, 11, 13, 34, 57, 70, 101, 102, 121, 122, 151, 154, 155, 178, 183

Alternation of turbulence 56

Average values of parameters 1, 3, 5, 6, 11, 13, 23, 26, 27, 31, 55, 56, 94–96, 113, 135, 139, 140, 163, 181

**B**

Bernoulli's equation 2, 19, 29, 34, 38, 40

Bifurcation 7

Blasius' problem 35

Boundary layers 2–4, 14, 22, 33, 34–36, 41, 51–53, 56, 57, 64, 65, 70, 82, 113–119, 122–124, 136, 138–142, 182, 183
- discontinuity 14, 51, 113
- instability 51, 57
- laminar 3, 34, 115, 138, 141
- separation 114, 115, 116, 118
- thickness 37, 70, 82, 113, 115, 122, 123, 138, 139, 141, 142
- turbulent 138, 139, 140

velocity profile 38, 51, 114, 119, 141
Breakdown threshold 173, 175, 177, 178

**C**
Chapman, Jouguet 126
Circulation velocity 20, 33
Conservation 3, 52, 138, 139
Combustion 103, 104
detonation 125, 126, 129, 130, 131, 143, 150, 151, 152–154
slow 103, 125, 138, 139, 144, 146, 147, 150, 151, 153, 154
zone 149, 153, 154
Compressed mixture 148, 149
Compressible flow 23, 24
Compression wave 53, 103, 107, 113, 115, 116, 126, 129, 139, 144, 153
Conversion into turbulence 3, 138, 139

**D**
Damping 2, 9, 14, 48, 52, 57, 73, 82, 84, 92, 93–96, 111, 118, 136, 152, 158, 183
Delayed potential 22, 35
Detonation (see: Combustion)
in pipe 125, 126
propagation 126
pulsating 152
reflection 129, 130, 133, 136, 151, 152, 154
spinning 152
transition to 125, 126, 127, 129–131, 144, 147, 150–154
Diffraction divergence 14, 48, 156, 157, 158, 160
Dirac delta function 23, 45
Discharge 155, 157, 160, 161–167, 170, 171–179
breakdown 161, 163, 164, 166–168, 170–178, 184
Discontinuity 13
boundary conditions for 27
compression 27, 47
simple wave 27, 47
tangential 13, 14, 51, 52
instability of 51
trajectory of ray 63
Dispersion relation 59
Displacement thickness 37, 38, 65, 86
Dissipation at wave reflection 69, 114, 115, 117, 154
Double saltus 126, 129, 130, 136–139, 141–144, 146, 149, 150, 152–154
Drag force 1, 2, 4, 11, 12, 22, 25, 29, 30, 35, 36, 38

**E**
Eddies 12, 13, 26, 182, 183
Energy conversion 52
Energy dissipation 6, 12, 24, 27, 40, 43, 49, 50, 52, 69, 92, 102, 182, 184
Energy flux 6, 49, 70, 101
Enthalpy 18, 29
Entropy 18, 27, 47
Euler's equation 19, 20
for incompressible flow 17
for a sound wave 253
for a viscous fluid 44
Exciting motion 1–3, 7, 8, 10, 11, 14, 15, 43, 51, 56, 87
Explosion 125, 150, 153

**F**
Flame 14, 103, 104, 106, 107, 135
acceleration 104, 125–127, 131–133, 136, 137, 141–144, 146
front 103, 104, 107, 108, 112, 113, 125, 129, 131, 133–139, 144
leading point 142, 149
normal velocity 107, 125, 126, 129, 131, 133–135, 146
piston analogy 103, 107, 108, 112, 142
propagation 103, 104, 108, 113, 125, 127–131, 133–137, 139, 141, 144, 146, 151
structure 104, 106–108, 112, 113, 125, 131, 134, 136, 137, 140–145
turbulent 136, 139, 154
wave emission 107, 125, 129, 130, 136, 138, 141, 144, 153
velocity oscillations 131, 136, 139, 141, 154

Fluctuations 1, 2, 4–8, 11,13, 14, 33, 42, 52, 57, 69, 87, 102, 118, 119, 122, 124
Fluid flow
around an angle 182
eddy flow 13, 26
ideal 18, 22, 24, 27, 29, 156
one-dimensional 8, 44, 46–48, 55
potential 2, 4, 14, 19, 21, 23, 26, 29, 33, 41, 57, 64, 68, 118
steady 7, 10–12, 19, 25–27, 29, 31, 33, 48, 58, 79, 82, 92, 94, 95, 136, 182
Taylor–Couette of 7
turbulent 1–7, 11, 13, 14, 31, 32, 56, 57, 63, 65, 66, 68, 69, 87, 101, 102, 136, 138, 155, 181–184
two-dimensional 2, 8, 33, 57, 75, 76, 89, 116
viscous 2–4, 11–13, 17, 21, 26, 27, 29, 31, 33, 38, 40, 43, 48, 51, 52, 57, 69, 70
Fourier transform 10, 48, 90–92, 95, 101, 155, 169, 170, 184

**G**
Geometrical acoustics 70
in moving media 58–61, 66
Green's function 24
Group velocity 59

**H**
Hugoniot adiabat 126
Hydrodynamic equations 1, 3–5, 8, 11, 13, 17, 18, 34, 102

**I**
Incompressible flow 11, 21, 22, 23, 25, 26, 27, 29, 33, 43, 63
conditions for 22, 23, 26, 27
Ionization potential 171, 172

**J**
Jet 7, 8, 13, 125, 182

**K**
Kármán constant 31
Kelvin's theorem on circulation 4, 14, 20, 33, 57
Kolmogorov's constant 6
Kolmogorov–Oboukhov concept 5, 6, 101

**L**
Laminar flow 1–3, 8, 9, 11, 31, 32, 34, 63, 65, 66, 68, 69, 115, 136, 138, 183
Landau constant 10
Laplace equation 13, 22, 29

**M**
Mach
number 25, 60, 121
reflection 179
Method of small perturbations 1–3, 8–10, 23, 26
Momentum flux 3, 11–13, 27, 30, 31, 48

**N**
Navier–Stokes equation 1, 3, 4, 6–8, 17, 18, 21, 22, 24, 43, 102, 181
Neutral curve 9
Nozzle 7, 8, 182

**O**
Odd–Zommerfeld equation 8, 9
Oscillations 7, 9, 103
acoustic at entry of channel 7, 8
disordered 87, 94, 96, 101
chaotization 82, 90–98
distribution in flow section 122–124
energy 101
hydrodynamic parameters 7–9, 14, 42, 46, 49–52, 56–59, 73, 78, 84–86, 88, 89–98, 100–103, 107, 113, 118, 120–124, 155, 168, 170, 180–184
initial arbitrary phase β 9
in sound wave packet 56
ordered 78
pendulum analogy 13
spectrum 89–94, 96–100, 102
structure 78, 95, 99, 101
superposition 13, 100, 102, 104, 107, 113, 118–120, 123, 167, 168, 183

**P**

Piston 43
  motion law 44
  motion irregularity 45
  motion spectra 44, 45
Poincare mapping 7, 120
Poisson's equation 13, 21, 22
Prandtl's equations 34
Pressure in flow cross-section 2–4, 29, 30, 33, 35, 38–41

**Q**

Q-curve 126

**R**

Rayleigh length 49, 157
Reynolds equation 4, 5
Reynolds criterion (number) 4–6, 9–12, 32, 39, 40, 41, 63, 107, 113, 122, 135, 139, 181
  critical value 4, 9, 63, 135, 136, 138, 141
  critical length 138, 141

**S**

Simple wave 13, 14, 43, 49, 53, 79, 80
  one-dimensional 43–48
  structure 22, 47, 51–53, 103, 104, 106, 107, 112, 113, 118, 132, 133, 136, 137, 140, 144
Slit scan 108, 110, 116, 127–129
Shock wave 47, 108, 114–116, 118
  oblique 114–116, 118, 126, 129, 138, 144–146, 149–154
  reflection from a solid surface 88, 90, 92, 93, 109, 110, 111, 114, 115–118
Spatio-temporal field 2, 4, 7, 8, 11, 13, 14, 15, 24, 26, 27, 52, 55, 56, 70, 73, 84, 85, 87, 93, 94, 96, 100, 102, 103, 113, 118, 119, 155, 167, 183, 184
Strange attractor 7
Stream function 8, 35–37, 40
Stream lines 19, 21, 33, 40

**T**

Tensor
  momentum flux density of 5
  turbulent stresses of 11
  viscous strain of 11
Turbulence
  energy dissipation 6, 12, 24, 27, 40, 43
  fully developed 56
  intermittency 56, 101, 183
  inverse problem 15, 155, 160, 174, 180, 184
  scale of 6, 7, 30, 81, 82, 101
  time spectral representation 97–99
  space spectral representation 97–99
  wave approach to 2, 3, 13–14

**V**

Velocity
  complex amplitude 9
  components 21, 27, 32, 35, 36
  gradient 33
  profile in flow 12, 21, 30, 32
  logarithmic profile 31
  perturbation propagation of 9, 22, 23
  potential of 23, 29, 55, 56
  vorticity 33
Viscosity
  dynamic 3, 17
  friction 2, 4, 12, 26, 38, 40, 48, 51, 52, 57, 69, 182
  kinematic 5, 21, 57, 138
Viscous force 2, 12
Viscous sublayer 69, 70

**W**

Wave
  amplitude 2, 8, 9, 10, 23, 24, 25, 55, 59, 67, 79, 86, 87–89, 95, 107, 111, 113, 124, 126
  beam of Gaussian profile 14, 48, 49, 156, 158
  beam of Besselian profile 14, 155, 156, 157, 160, 161–164, 167–173, 175–180, 184

damping of 14, 52, 65, 67, 69, 85
distribution of parameters in 13, 14, 24, 55, 56, 85, 89–91, 155, 183
energy conversion into heat 52
energy dissipation 43, 49, 52
equation 8, 13, 55, 85, 87, 100
equation general solution 23, 24
front shape 55, 56, 58, 59, 61, 62, 65, 66, 68, 73, 75–78, 92, 110, 114, 116, 154, 157, 171, 183
length 48
number 48
probe point trajectory 59–70, 116, 117, 119–123, 176
train 46, 99, 175
vector 48

GPSR Compliance

*The European Union's (EU) General Product Safety Regulation (GPSR) is a set of rules that requires consumer products to be safe and our obligations to ensure this.*

*If you have any concerns about our products, you can contact us on ProductSafety@springernature.com*

In case Publisher is established outside the EU, the EU authorized representative is:

Springer Nature Customer Service Center GmbH
Europaplatz 3
69115 Heidelberg, Germany

**Batch number: 10583998**

Printed by Printforce, the Netherlands